MODERN BATTERIES

MODERN BATTERIES

An Introduction to Electrochemical Power Sources

Second Edition

C A Vincent
School of Chemistry,
University of St Andrews, UK

B Scrosati
Dipartimento di Chimica,
Universita degli Studi di Roma 'La Sapienza', Italy

BUTTERWORTH
HEINEMANN

OXFORD AMSTERDAM BOSTON LONDON NEW YORK PARIS
SAN DIEGO SAN FRANCISCO SINGAPORE SYDNEY TOKYO

Butterworth-Heinemann
An imprint of Elsevier Science
Linacre House, Jordan Hill, Oxford OX2 8DP
200 Wheeler Road, Burlington, MA 01803

First published 1984
Second edition 1997
Transferred to digital printing 2003

British Library Cataloguing in Publication Data
A catalogue record for this book is available from the British Library

Library of Congress Cataloguing in Publication Data
A catalogue record for this book is available from the Library of Congress

ISBN 0 340 66278 6

For information on all Butterworth-Heinemann publications
visit our website at www.bh.com

Contents

Preface to the second edition

Much has happened in the world of batteries since the preface to the first edition was written almost 15 years ago. Some of the developments were predicted at that time, some were not. Perhaps the most important factor, which has led both to a renaissance in new developments and to a significant growth in demand, has been the phenomenal expansion in microelectronics-based, high value consumer products which need secondary cells with excellent energy density, good charge retention and other demanding electrical characteristics. The 'three Cs' – cellular telephones, portable computers and camcorders – typify such applications.

So far as systems are concerned, one of the most significant commercial successes in this period has been the emergence of the nickel–metal hydride cell, which is now widely available for consumer products and is being evaluated for traction applications. The elimination of cadmium (and that of mercury from primary zinc cells) is seen as a very important contribution to protection of the environment. A more recent development has been the major commercialization of rechargeable alkaline zinc cells, which in the next decade may take over a significant part of the market currently occupied by primary batteries. Perhaps the most important advances, however, have been in the area of rechargeable lithium batteries. 'Lithium ion' cells have now been in the market for 6 years, and production is estimated to be 10 million cells per month by 1997. Polymer-based lithium cells have taken much longer to develop to a commercial level, but it is expected that they too will become a significant market sector by the end of the decade. In the meantime, development of the more traditional aqueous cells has not stood still. Nowhere is this more obvious than in the advances in sealed and valve regulated lead–acid (VRLA) batteries. One application which has been long predicted but slow in realization has been the application of batteries as power sources in electric vehicles. Legislation in California and elsewhere in the USA, together with significant investment there and in Europe and Japan, has, however, now brought commercialization closer. Development is proceeding both in aqueous systems and in advanced batteries, where the Zebra sodium–nickel chloride cell and a number of lithium-based systems are in competition.

The passage of time has also taken its toll at a personal level, and it was with great sadness that we learned of the death of our co-author and close friend Mario Lazzari at an early age in 1986. Our other co-author, Franco Bonino, was unable to contribute to this edition, and we were therefore delighted that Kathryn Bullock (AT&T Bell Laboratories) agreed to update the chapter on lead–acid batteries, and that Jim Sudworth (Beta Research & Development Ltd) offered to completely rewrite the chapter on high temperature systems. We had also invited another colleague, Dr Sid Megahed, to help us with the chapter on rechargeable alkaline cells, but his sudden and much regretted death, also at an early age, deprived us of this contribution. We are particularly indebted to Professor Karl Kordesch at the Technical University Gratz, the doyen of the alkaline battery field, who at very short notice was unstinting in his provision of assistance and constructive criticism for sections of this chapter. As in 1982, we must again thank our many friends in industry and in research laboratories for their help in writing this book, and the battery companies who have been so generous in their provision of technical information, photographs and diagrams. Finally we would like to thank Gladys MacArthur of the University of St Andrews for typing the manuscript and coordinating all the work.

Our aim has been to provide a brief but reasonably comprehensive book for the non-specialist. With the exception of the first two chapters, most of the book has been rewritten and rearranged to justify the title. We hope that it will be of value not only to electrochemists, but to anyone interested in the field of chemistry, materials, electrical engineering and technology in general.

St Andrews
March 1997

Preface to the first edition

Electrochemical power sources range from miniature batteries with an energy storage capability of less than 0.1 Wh to projected load levelling modules with capacities greater than 10 MWh. All such systems utilize the energy evolved by spontaneous chemical reactions to produce electric power directly, and all rely on the same fundamental physical processes for their operation. Commercial batteries have now been manufactured for over a century, but until comparatively recently, research and development in the battery industry has been directed largely towards improvements in well tried systems, especially in the fields of engineering design and production. However recent advances in electrochemistry and materials science have opened the way for the evolution of entirely new types of power source with greatly improved electrical performance and other desirable characteristics.

Our aim in writing this book has been to produce a brief but comprehensive account which may be read – we hope with some profit and interest – by anyone with a basic knowledge of chemistry and physics who wishes to know something about how batteries work, and what the main developments are in this fascinating area of science and technology. We have tried to maintain a balance between describing well established 'conventional' systems, and 'state-of-the-art' developments which may or may not become of commercial importance. However, because of the existence of excellent specialized texts which describe in some detail the evolution of the main commercial batteries, we have placed considerable emphasis on discussing recent trends and discoveries.

As the book has been written for the non-specialist, the theoretical background to the basic processes involved in cell operation is described in some detail in preference to a more thorough series of comparisons of the characteristics and performance of competing systems. We have excluded any discussion on the very closely related field of fuel cells since a number of accounts of this topic have been published recently. It has been our intention to describe and characterize most of the established and emerging primary and secondary battery systems which are of current commercial or theoretical interest. Research into novel power sources may shortly lead to the major breakthroughs necessary before electric vehicles become a major component of the transportation system, and electrochemical storage becomes a serious competitor for large-scale load levelling applications. It is our hope that this book may lead to a more general appreciation of the role of the electrochemical power source today and its potentialities in the future.

We have to thank our many friends both in industry and in research laboratories for their assistance in the preparation of this book. In particular we would like to acknowledge Dr J Thompson (Royal Aircraft Establishment (MOD), Farnborough) who reviewed the whole manuscript, and the following colleagues who gave us valuable advice: Dr P Bruni (Superpila), Dr W G Bugden (British Rail), Dr Ing G Clerici (Magneti Marelli), Dr M D Ingram (University of Aberdeen), Dr R Marassi (University of Camerino), Dr B B Owens (Medtronic) and Dr B Rivolta (Polytechnic of Milan). We would also wish to thank the battery companies from all corners of the globe who were so generous in their provision of technical information, photographs and diagrams. Finally we thank Ann Hughes of the University of St Andrews for her patience in typing and re-typing the manuscript.

It may be considered appropriate to mention the Scottish/Italian collaboration which has resulted in the writing of this book. It started in 1965 when two of us (BS and CAV) were working in the laboratories of H.A. Laitinen at the University of Illinois and expanded to include members of the CNR Centre on Electrode Processes at Milan when a joint research programme was instituted some 10 years ago. We are pleased to acknowledge the financial assistance given by the Science and Engineering Research Council (UK), the Consiglio Nazionale delle Ricerche (Italy), NATO and the British Council to foster this collaborative venture over the years.

St Andrews
June 1982

1 Introduction

Colin A Vincent

1.1 Electrochemical power sources

An electrochemical power source or battery is a device which enables the energy liberated in a chemical reaction to be converted directly into electricity. Batteries fulfil two main functions. First and foremost they act as portable sources of electric power. Well known modern examples range from the small button cells used in electric watches to the lead–acid batteries used for starting, lighting and ignition in vehicles with internal combustion engines. The second function, which is likely to increase in importance over the next 20 years, is based on the ability of certain electrochemical systems to store electrical energy supplied by an external source. Such batteries may be used for driving electric vehicles, for emergency power supplies, and as part of the main electricity supply system for meeting short duration demand peaks (load levelling) or in conjunction with renewable energy sources, such as solar, wave or wind power.

The first authenticated description of an electrochemical battery was given by Alessandro Volta, Professor of Natural Philosophy (Physics) at the University of Pavia in Italy, in a letter to the Royal Society (London) in 1800. A photograph of an original Volta 'pile' is shown in Fig. 1.1. The importance of Volta's discovery as a tool for advancing the understanding of chemistry and physics was immediately grasped by scientists in a number of countries. However, it was the introduction of telegraph systems, which were becoming of increasing importance in the 1830s, that gave rise to the development of reliable commercial batteries, capable of sustaining a substantial flow of current without undue loss of cell voltage.

The first high current electroplating battery was described in 1840, and over the next two decades the use of techniques such as electroplating and electroforming, together with the exploitation of practical electric motors, gradually became more widespread. In the 1870s, a more general consumer

Fig. 1.1 Volta pile

market for batteries was created by the manufacture of electric bell circuits for homes, offices and hotels. The 'flashlight' was introduced at the turn of the century, some 20 years after Edison's invention of the incandescent lamp. By then the annual production of batteries in the USA alone had exceeded two million units.

The large-scale introduction from 1870 onwards of dynamos or electro-magnetic generators driven by heat engines led to the worldwide industrial and domestic use of 'mains' electricity, accepted as commonplace today, a century later. This ready availability of electrical energy was the principal motivation for the development of secondary or storage batteries, although this area was soon to be further stimulated by the demands of the growing motor car industry.

A further impetus for commercial battery development came with the introduction of domestic radio receivers in the 1920s, and an equivalent growth has been seen over the last 30 years with the development of microelectronics-based equipment. Today, it is estimated that annual battery production totals 8–15 units per head of population throughout the developed countries of the world.

The world market for batteries of all types now exceeds £100 billion. Over half of this sum is accounted for by lead–acid batteries – mainly for vehicle starting, lighting and ignition (SLI), and industrial use including traction and standby power, with about one-third being devoted to primary cells and the remainder to alkaline rechargeable and specialist batteries.

1.2 Nomenclature

There is some confusion in the terminology used to denote the electrochemical devices which convert chemical into electrical energy. In many cases, the devices have changed in character with the passage of time, but have retained their original names; in others, the terms commonly used do not clearly define the nature of the device. In this book we conform to general usage and employ the words 'cell' and 'battery' interchangeably to describe a closed electrochemical power source, i.e. one in which the reactants are incorporated during manufacture. The term 'battery' originally implied a group of 'cells' in a series or parallel arrangement, but is now understood to mean either a single cell or a group of cells.

Two other terms which are not self-explanatory are 'primary' and 'secondary' cells. A primary system is one whose useful life is ended once its reactants have been consumed by the discharge process. In contrast, a secondary system is capable of being charged or recharged when its reactants have been used up: the spontaneous electrochemical reaction can be reversed by passing current through the cell in the opposite direction to that of cell discharge. A secondary battery might therefore be considered as an electrochemical energy storage unit. Note, however, that the energy derived from the external current is stored as chemical energy, and not as electrical energy as in a capacitor. Other terms are sometimes used to describe this system, e.g. accumulator (introduced by Davy along with the terms 'cell' and 'circuit'), storage battery, rechargeable battery, etc. In a reserve cell, one component (usually the electrolyte) is separated from the rest of the battery or maintained in an inactive condition until the cell is activated. Such cells are capable of very long-term storage in environmentally hostile conditions since self-discharge and other chemical processes are minimized. Reserve cells are used in applications such as life-jacket or life-raft lights, or in missile weapon systems.

We do not consider the related subject of fuel cells, where both cathodic and anodic reagents – usually gases – are stored externally and can be supplied to the electrochemical cell on a continuous basis. A number of books have recently been published on this topic. The term 'hybrid cell' is used here to describe a power source in which one of the active reagents is in the gaseous state, e.g. the oxygen of the air. Use of the word 'hybrid' in this context should not be confused with its meaning in the phrase 'hybrid electric vehicle', which refers to an electric vehicle with more than one power supply, as described below.

A large number of technical terms are associated with the literature on batteries: the more common of these are given in the Glossary, while the electrical quantities used to describe battery performance and characteristics are defined in Section 2.5, and summarized in Appendix 4.

The most common convention for writing an electrochemical cell is to place the negative electrode on the left and the positive on the right. The cell is then named in the same way: thus reference to the 'sodium–sulphur cell' implies that sodium is the active reagent at the negative electrode and sulphur is that at the positive electrode. We make three exceptions to this general rule in order to conform to normal usage, and call the lead–lead oxide cell the 'lead–acid cell', the cadmium–nickel oxide cell the 'nickel–cadmium cell', and the zinc–manganese dioxide cell the 'Leclanché cell'.

1.3   The renaissance in battery development

Until very recently, 'conventional' batteries using solid electrodes and aqueous electrolytes proved satisfactory for the majority of common applications. Traditional primary systems, such as the Leclanché Zn-MnO_2 cell and the alkaline manganese cell, have been (and to a large extent still are) adequate power sources for portable electrical equipment. Well established rechargeable batteries such as those based on the lead–acid or nickel–cadmium systems have for a long time been employed as small localized energy storage units (e.g. in rural areas, in railway and telephone systems, etc.) and as sources of auxiliary power in ground, air and sea transport. For many years, research and development in the battery industry has been directed mainly towards improvements in these well known systems, especially in the fields of engineering design and production.

In the past 25 years, however, the situation has changed considerably. First, advances in semiconductor technology have led to the production of large-scale integrated (LSI, VLSI and ULSI) circuits in immense numbers, bringing about a revolution in the electronics industry. Microelectronic components are now inexpensive and are widely used in the production of pocket calculators, electric watches and similar devices. In 1990, the world production of battery-powered watches was 4×10^8. Development of a wide variety of such electronics-based consumer products soon demanded the evolution of miniature power supplies which would offer a much higher energy per unit volume and superior discharge characteristics as compared with those of traditional batteries.

The second, and perhaps more important, factor affecting the demand for new battery systems was the realization in the late 1960s that the constantly increasing energy needs of the developed countries of the world would soon lead to the progressive exhaustion of oil supplies. This in turn led to the requirement for more efficient use of the remaining fossil fuel reserves and for a shift towards the exploitation of alternative energy sources, preferably of a clean and regenerative type. Central to the problem both of the utilization of discontinuous energy sources, e.g. solar, wind and wave power, and of the efficient running of conventional generating plant is the provision of suitable energy storage systems. While there are a number of alternatives to be considered, such as pumped hydroelectric or compressed air storage, electrochemical storage batteries are in many instances more convenient, being transportable and flexible in size, as well as being silent and non-polluting. For this application, batteries require the ability to undergo large numbers of deep charge/discharge cycles with high efficiency and without physical degradation.

Since a considerable proportion of all petroleum is consumed in vehicle traction – a particularly inefficient way of extracting energy from a scarce resource which simultaneously causes severe environmental pollution in urban areas – the possibility of replacing vehicles driven by internal combustion engines with battery-powered electric transport is under active consideration, and the development of advanced batteries for this purpose is being pursued in a number of countries. Since batteries for electric vehicles (EVs) must be transported as part of the vehicle load, they require high power/mass ratios in addition to high cycle efficiency.

1.4   A survey of common battery types and applications

The total available energy of a battery is a measure of how much electricity it can deliver (usually measured in Wh) and is directly related to the size of the unit. One possible classification of the most common commercial battery types, according to size, is given in Table 1.1. The range of battery energies extends over at least 15 orders of magnitude. At the bottom of the range there are 0.1 cm^2 experimental cells with a PbF_2 solid electrolyte which have a total energy of just over 1×10^{-6} Wh. The smallest commercial button cells have energies of around 100 mWh, while the common 'D size' cylindrical cells, which have a total volume of 45 cm^3, have energies in the range of 2–15 Wh. Rechargeable cells used in power tools and other 'cordless' electric appliances can supply 20–100 Wh. At the top of the range, submarine lead–acid batteries weighing nearly 200 tonnes have rated energies of 3 MWh, while the load levelling batteries of 40 MWh have now been realized.

Miniature batteries

Miniature batteries based on aqueous, non-aqueous and solid electrolytes are manufactured as power sources for microelectronics and other miniaturized equipment. In Fig. 1.2, the sizes and shapes of some representative button cells are shown. A typical application for such cells is in the electric watch, where the oscillator circuit draws a continuous current of 0.2–0.6 µA and depending on the type of frequency divider and display, the complete unit may require a total of up to 0.5–2.0 µA for operation. Hence the total amount of electrical energy consumed in driving the watch for a year is in the range 15–60 mWh. At present, batteries are manufactured which last for 5–10 years. Watch batteries must have exceptionally low self-discharge rates and very reliable seals to prevent leakage. Further, they

Table 1.1 Classification of batteries according to size

Type	Energy	Applications
Miniature batteries	100 mWh–2 Wh	Electric watches, calculators, implanted medical devices
Batteries for portable equipment	2 Wh–100 Wh	Flashlights, toys, power tools, portable radio and television, mobile phones, camcorders, lap-top computers
SLI batteries (starting, lighting and ignition)	100–600 Wh	Cars, trucks, buses, tractors, lawn mower traction
Vehicle traction batteries	20–630 kWh (3 MWh)	Fork-lift trucks, milk floats, locomotives (submarines)
Stationary batteries	250 Wh–5 MWh	Emergency power supplies, local energy storage, remote relay stations
Load levelling batteries	5–100 MWh	Spinning reserve, peak shaving, load levelling

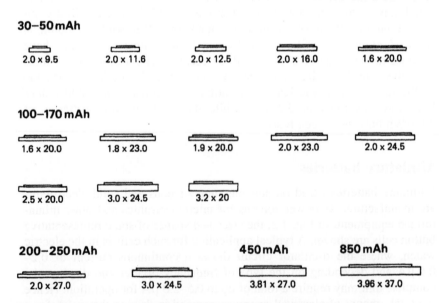

Fig. 1.2 Dimensions (in mm) and capacities of some representative button cells

have strict design constraints, in order to fit into the restricted available space within the watch case. From an electrical point of view, as well as having a high volumetric energy density (i.e. the amount of energy delivered per unit volume) the battery must display a flat discharge characteristic for accurate operation of the watch circuits. In addition, many watch batteries must be able to sustain occasional high rate discharge pulses, since watches with liquid crystal diode displays usually have a small tungsten lamp for back lighting. An alternative approach for the provision of power for watches is to use a miniature battery of relatively small capacity which can be recharged using solar cells.

An excellent example of the effects of the simultaneous developments of microelectronic circuitry and battery technology has been the increase in service life and reduction in physical size of cardiac pacemakers. The earliest pacemakers, implanted in the 1960s, were simple pulse generators operating at fixed rates, for periods of 1–3 years. They were powered by zinc–mercury oxide batteries, and the whole unit weighed about 200 g. With the development of the high energy density lithium–iodine solid state cell, together with further miniaturization of the electronic circuits, modern 'demand' pacemakers, whose pulsing is inhibited by physiological signals indicating normal ventricular activity, now weigh under 50 g and have a lifetime of 10 years. Typical cells have a total volume of 6 cm^3 and a mass of 22 g.

Demand pacemakers are very low current devices, requiring only 25–50 μW for sensing and 60–100 μW for stimulation. In contrast, implanted ventricular defibrillators (Fig. 1.3) must be able to deliver short electric pulses of 25–40 J (e.g. 2 A at 2 V for 10 s) which can shock the heart into normal rhythm, and hence require a much higher rate battery. The most common system is a lithium–silver vanadium oxide cell with a liquid–organic based electrolyte. More than 80 000 such units have been implanted. Implanted drug delivery devices also use lithium primary batteries, as do neurostimulators and bone growth stimulators.

Batteries for portable equipment

Perhaps the best known of all batteries are the inexpensive packaged 'dry batteries' used in a wide variety of portable lighting appliances, battery operated toys, radios and numerous other applications. A number of standard sized cylindrical single cells are shown in Fig. 1.4. The majority of this class of battery are based on the Leclanché $Zn\text{-}MnO_2$ system. Although the original was introduced towards the end of the last century, there have been (and still are) notable and continuing improvements in the operating characteristics of this system. However, such batteries tend to have rather poor voltage regulation; i.e. the cell voltage decreases as discharge through a fixed load progresses. The introduction of the Ruben–Mallory $Zn\text{-}HgO$ cells in the 1940s provided a power source with a very flat discharge characteristic and an improved energy density. More recently, even higher

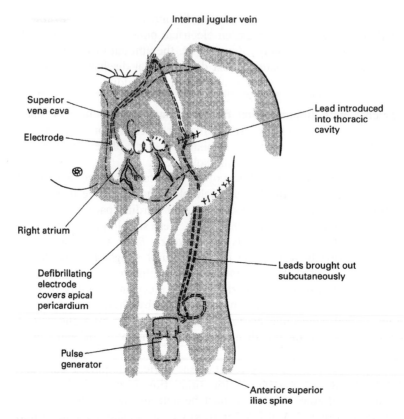

Internal jugular vein

Superior
vena cava

Electrode

Right atrium

Defibrillating
electrode
covers apical
pericardium

Pulse
generator

Lead introduced
into thoracic
cavity

Leads brought out
subcutaneously

Anterior superior
iliac spine

Fig. 1.3 Schematic illustration of the implantation of an automatic cardiac defibrillator. (By permission of *World Medicine*: J.H. Tanne, 1981, **16(25)**, 64.)

energy densities have been achieved by cells having lithium anodes and electrolytes based on organic solvents. Such batteries are relatively expensive compared with Leclanché batteries, but are used extensively where light weight is important.

For many applications where a relatively high current drain is involved, it is much more cost-effective to use a sealed or 'maintenance-free' rechargeable battery as an electrical energy store. This is true, for example, for 'cordless' power tools, portable television receivers, hedge trimmers and lawn edging tools, emergency lights, etc. Examples of such batteries are the alkaline nickel–cadmium, nickel–metal hydride, the alkaline iron–nickel oxide, the sealed lead–acid systems and, more recently, lithium ion batteries. Many of these batteries in the 10 mAh–15 Ah range have exactly the same external dimensions as the primary cells they are designed to replace. Larger units with capacities up to 1000 Ah are also available. The alkaline batteries have an excellent cycle life (i.e. they can undergo a large number of charge/discharge cycles), perform well at very low temperatures, have reasonably good voltage regulation and are very robust. They are, however, much more expensive than their lead–acid equivalents. Both

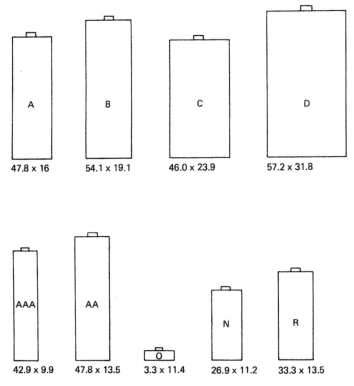

Fig. 1.4 Standard dimensions (in mm) for cylindrical primary cells

varieties suffer from poor energy density (i.e. amount of energy delivered per unit mass) and may be replaced in the future by lithium–organic systems for applications where light weight is desirable.

SLI batteries

Almost 250 million batteries for starting, lighting and ignition (SLI) in internal combustion engine vehicles are manufactured annually, using over one-third of the total world output of lead. Such batteries are required to 'turn' engines with high compression, often at low temperatures where the viscosity of engine oil is high – currents of up to 500 A may be drawn for this purpose. In addition, the battery must supply power for ignition, lighting, ventilation, rear window heaters, etc. Most modern car electric circuits employ a nominal 12 V system and the batteries have six lead–acid cells in series, with capacities of the order of 40–60 Ah. Significant improvements in materials and design have raised the energy density of SLI batteries to around 45 Wh/kg and 75 Wh/dm^3, while retaining durability and service life. (It is inappropriate to talk of 'cycle life' in this context since it is unusual for SLI batteries to undergo deep discharge.) The development of new lead alloys, separator envelopes and battery casings has led to the emergence of the 'maintenance-free' SLI battery which requires no (or rare) addition of water during its lifetime, excellent charge retention during storage and no significant terminal corrosion. Because of the large scale of their manufacture, SLI batteries are relatively inexpensive and are therefore often used quite successfully for other applications, e.g. to provide tractive power for lawn mowers and 'go-karts', in emergency lighting units, etc.

Nickel–cadmium batteries with thin sintered plates are used for on-board power supplies in aircraft, helicopters, tanks and military vehicles where their excellent low temperature, high rate performance is an important attribute. Modern 40 Ah cells designed for airborne use can deliver 20 kW of instantaneous power at 25°C and over 10 kW at −30°C. Again, the high cost of this system compared with that of lead–acid batteries has restricted its use.

Vehicle traction batteries

As pointed out above, the advantages of the widespread use of electric batteries for vehicle traction are numerous: in addition to improvements in the environment effected by the silent, pollution-free operation of EVs, conservation of oil supplies and a more efficient performance of the mains electricity generating system, brought about by load levelling due to overnight charging of EV batteries, would result. Electric vehicles have a long history, stretching back to the late 1830s when the world's first 'electric carriage' was built in Scotland. Before the end of the 19th century a French battery-powered car had pushed the world land speed record above 100 km/h (62 mph), and by 1916 annual production of electric vehicles in the USA had reached 10 000. EVs have fewer moving parts than other vehicles, and the inherent lack of vibration of the electric motor cuts down wear, reducing the need for maintenance and extending the vehicle's life. Electronic control and regenerative braking made EVs easy to drive, especially in heavy urban traffic. Despite the fact that the uncertified world speed record for an EV is now over 188 mph, there is currently no major production line for commercially competitive general passenger or freight vehicles. The reason for the relative lack of commercial success of electric road vehicles up to the present lies in the battery specifications required in order to produce a performance (and cost) approaching those of internal combustion powered vehicles. The speed, acceleration and range of most present-day EVs are limited by the low energy and power densities of their traction batteries.

A typical family car, having a laden weight of 1–1.5 tonnes would require 5–10 Wh of energy to move 50 km. Taking the lower figure, this would involve the consumption of approximately 4.5 dm^3 or 3.9 kg of petrol. An average value for the energy density of a lead–acid battery capable of deep cycling at a useful rate might be 25 Wh/kg. Hence a battery weighing some 200 kg and occupying 120 dm^3 would be required to drive the car the same distance. In other words, the lead–acid energy store is up to 50 times heavier and uses up 25 times the space of its petrol equivalent. Other disadvantages of EVs include the restricted power available for acceleration and hill-climbing and the time taken to recharge the battery. One might compare this latter period of 6–12 hours with the 2–3 minutes needed to refill a petrol tank. It is interesting to calculate that the equivalent power flow through a normal petrol pump delivery hose is about 30 MW!

As far as specialized and off-road vehicles are concerned, however, electric traction using rechargeable batteries forms a significant sector of the market. The materials handling industry makes extensive use of battery-powered vehicles, ranging from fork-lift trucks and turret trucks to various tractors. The noise-free, easy starting and simple operation of these vehicles using electronic controls, together with the absence of exhaust fumes, makes them ideal for use in enclosed areas such as warehouses and factories. A wide variety of tractors is needed for baggage, supplies and personnel ('people movers') in airports, hospitals and large industrial complexes. Electric golf buggies and electric wheelchairs are a common sight worldwide. The UK has one of the largest fleets of registered EVs. The reason for this is the existence of some 25 000 'milk floats', or bottled milk delivery trucks. The tradition of daily home delivery of milk in the UK is, however, declining (20% in the past 5 years) with competition from long-life milk in the supermarkets. This, combined with the longevity of these vehicles (>25 years), has resulted in a 10-fold reduction in the manufacture of new milk floats.

The future of battery driven vehicles for specialized purposes, such as the electric locomotives using during construction of the Channel Tunnel which used 100 kWh batteries, or the floor sweepers/scrubbers now used in airport terminals, industrial premises and railway stations, is assured. What is of much greater significance is the future of EVs in the general automobile industry which currently has a turnover of some £500 billion.

There are two reasons why battery-powered EVs are likely to become increasingly common over the next 50 years. The most important is that petroleum production from natural oil reserves has now peaked and will fall rapidly over this period. Production of synthetic fuels is expensive and is unlikely to be able to cope with demand. Electricity generation from coal, nuclear fission or renewable resources is assured for a considerably longer period, so that energy for EVs will be readily available. The second influence on the eventual replacement of internal combustion engine vehicles by EVs is the growing concern over polluting emissions and the desire to live in an environmentally responsible society. In 1990 the state government of California promulgated regulations requiring 2% of all cars sold in the state by 1998 to be 'zero emission' vehicles, and that this figure should be increased to 10% by 2003. These goals were later seen to be slightly overambitious and the timescale has been relaxed. Despite this, there is no question that the programme will be cut back, and the example of

California is being followed by other states, by European countries and by Japan. In 1992 the first contracts were announced for a $260 million collaborative research programme by the United States Advanced Battery Consortium (USABC). In Japan, five battery companies together with the electric power industry and other manufacturers have been brought together by MITI in a $100 million programme known as LIBES to develop secondary lithium batteries. The objectives of these research programmes are to develop and demonstrate the feasibility of manufacturing advanced batteries which will have three times the energy output in comparison with the same weight of current lead–acid traction batteries, while simultaneously providing high power and high cycle life. This initiative is leading to the development of new battery systems at a pace which has been compared with that of the manned space programmes of the 1960s.

There will probably be a number of stages in the evolution of EV batteries. In the short term, 'advanced' versions of existing battery systems will be widely used. Prototype lead–acid traction batteries with energy densities in the range 40–60 Wh/kg have been described by several manufacturers; it seems unlikely, however, that batteries of this type which are capable of more than 1000 deep cycles will ever have energy densities much above 40 Wh/kg. Nickel–cadmium systems are unlikely to make a significant contribution, due to the toxicity of cadmium and also to its relative scarcity. A closely related system, however, nickel–metal hydride is likely to fulfil midterm objectives with energy density in the range 80–100 Wh/kg and good recharging characteristics. The high cost of nickel and problems of self-discharge are significant drawbacks. Other ambient-temperature aqueous electrolyte batteries being considered for EV use include zinc–nickel oxide (75 Wh/kg), zinc–air (130 Wh/kg) and zinc–chlorine hydrate (80 Wh/kg) systems. In the longer term, advanced batteries using sodium or lithium with energy densities of over 100 Wh/kg are likely to form the basis for EV traction. Sodium-based cells working at high temperature (250–370°C) have been produced at a pilot-plant level and road tested in a variety of vehicles. High temperature lithium batteries using molten salt electrolytes have also been under development for many years. The most promising long-term solution is the ambient (or relatively low temperature) lithium ion cell using an immobilized or polymer-based electrolyte. Such batteries are now under intensive development worldwide.

A hybrid electric vehicle has more than one type of power supply to support the drive. These may be used independently, in series or in parallel to increase the range of the vehicle or to improve its overall efficiency, for example by recovering energy by means of regenerative braking. The combinations of battery plus mains power and battery plus internal combustion engine have proved the most promising examples of practical hybrid EV power supplies, but the increased complexity and cost of these vehicles has prevented their commercial development so far. Battery/battery hybrids have also been tested: for example, a high performance, but heavy, lead–acid battery may be used to provide power for acceleration and hill-climbing, while a light zinc–air battery with high energy density supplies the energy for running at constant speed.

Battery-powered trains or railcars have been in operation since the turn of the century, mainly in Germany. They have the advantage over electric road vehicles of much lower frictional losses, so that energy density limitations are not so severe. Modern railcars seating 80 passengers and operating on 630 kWh lead–acid batteries (which have a mass of 21 tonnes and a cycle life of 1250–1550) can accelerate at 0.4 m/s^2 to 100 km/h on the level, and have a range of 250–450 km. High reliability, low maintenance costs and safety are the important attributes of this form of transport. The use of hybrid mains/battery railcars is currently being investigated. For example, a relatively small 1 tonne sodium–sulphur battery would allow a railcar to make a circuit of some 100 km away from an electrified line.

Submarines of the non-nuclear reactor type rely on storage batteries for propulsion in the submerged condition. Generally the lead–acid system is used, but the light zinc–silver oxide batteries have been installed in some vessels. The energy stored is in the range of 2–3 MWh, and the batteries can weigh more than 180 tonnes. Special precautions are required for such large units to ensure adequate heat dissipation and thus to prevent the build-up of excessive temperature differentials.

There is also a wide range of more unusual vehicles which have been developed to use battery traction. For example, passenger vessels to work on inshore waters are being developed in Germany which are driven using batteries charged by photovoltaic cells. A motorcycle using zinc–silver oxide batteries has travelled at over 165 mph, while an electric 'dragster' has covered a standing quarter mile in 11 seconds. Battery-powered aircraft have been experimented with since the early 1970s. The most successful manned machines have been solar/battery hybrid-powered sailplanes and battery-powered ultralights. A more serious development is that of unmanned aircraft. For a number of years such craft have been used for both military and civilian remote surveillance and reconnaissance: a payload of several kg may be combined with a range of 50–100 km and a flight duration of 1 hour. Lithium–sulphur dioxide primary cells are proving to be an ideal power source for this application. The advantages to military users of a vibration-free platform with low noise and zero infrared emission are obvious. Now, consideration is being given to the development of lightweight remotely controlled aircraft for other more sinister aspects of warfare.

Stationary batteries

The main functions of storage batteries in the 250 Wh–5 MWh range are to provide standby power for emergency lighting, telephone services, hospital equipment, etc., or for local storage in remote relay stations, radio beacons, navigation buoys, etc. In both types of application it is rare for deep cycling to be involved. Generally a continuous 'trickle' charge is available (see Appendix 1), either from mains power or from solar, wind or wave energy. Stationary batteries for standby power are usually constructed of a special design of lead–acid cell known as the 'high performance Planté' type. This form of battery generally has very low standing losses and very long lifetimes; 20–30 years are typical. About 25% of standby power is based on nickel–cadmium batteries.

A schematic circuit for an emergency power supply is shown in Fig. 1.5. The load is fed directly from the mains supply (which also 'float' charges the battery), but in the event of mains failure, the load is automatically switched to the inverter by the mains monitoring system.

Windmill/battery supplies for automatic telephone exchanges have been successfully implemented in remote locations in Denmark using 400 Ah lead–acid batteries. Small wavepower air turbines ($\approx$100 W) have been used to charge batteries on maritime navigation buoys since 1965, and solar energy/battery systems have worked well in survey buoys, for supplying energy for light beacons near airports, for remote road warning signs, for telemetry systems and for a wide range of similar applications. Spacecraft batteries which are recharged by solar energy are shown in Figs 6.10 and 6.23.

Batteries for load levelling

A schematic smoothed national electricity demand curve is shown in Fig 1.6. Such a graph invariably shows a deep minimum during the night, followed by increased demand during waking hours. Superimposed on this curve are a number of sharp spikes, usually associated with cooking and television watching habits. The electricity generating industry must therefore maintain plant capable of producing in excess of the peak power requirement. This is generally accomplished by generating:

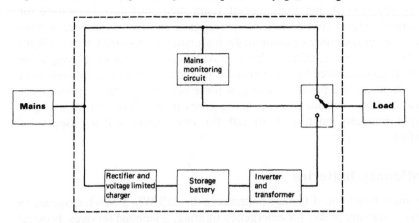

Fig. 1.5 Schematic circuit for uninterruptible emergency power supply

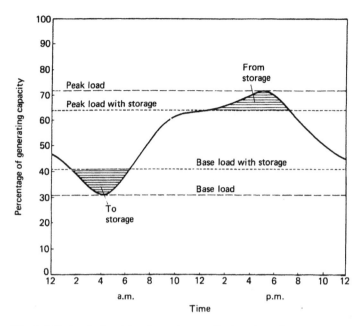

Fig. 1.6 National electricity demand curve. In the winter the curve moves up the *y*-axis. In the UK, the winter demand is roughly double that of summer

- the base load with nuclear reactors or other efficient generators running at their optimum capacity;

- the intermediate load, by adding older less efficient plant, normally at part load;

- peak demand, by bringing onto line expensive gas turbine-driven generators.

The principle of load levelling is to increase the base load capacity and to use the excess capacity available during the demand trough to stockpile energy which may be used to meet the peak demand later in the cycle. A number of techniques are currently employed to put load levelling into effect. In mountainous temperate countries pumped hydro-storage is economical despite high capital costs. Individual schemes of 200–2000 MWh capacity have been running for many years throughout the world. Other methods include thermal storage, compressed air storage (in combination with a modified combustion turbine), hydrogen storage, flywheels, superconducting magnetic energy storage (SMES), etc.

Batteries are likely to find an increasing application in this role. Although electrochemical storage may prove a less cost-effective alternative in some situations, it has many advantages. Batteries have a much shorter lead time in manufacture than most competitive systems, and being modular (unlike a hydroelectric dam or compressed air store), the energy storage facility can be added to, split into smaller units, or even transported to a new site. A particular advantage is that small units, free from environmental problems, can be situated in urban locations near industrial load centres, and so lead to very significant savings in transmission and distribution costs. This is especially important if the distribution network is reaching its maximum rated capacity during peak demand. Under these circumstances, it may be much less costly to install a battery storage system at the load centre, rather than to disrupt a large urban area in order to install extra distribution cables.

In addition to load levelling between night and day, the fast response of batteries to energy demand may enable them to contribute to two other areas of grid power supply: (1) for peak shaving, i.e. for meeting short duration regular or predictable demand peaks; and (2) to act in place of spinning reserve. 'Spinning reserve' refers to the practice of maintaining a number of generating plants running at rates below their maximum efficiency in order to be able to meet any sudden unexpected demand, say with a 10 second response time.

In the USA, the Electric Power Research Institute (EPRI) and the Energy Research and Development Administration (ERDA), now the US Department of Energy (DOE), established a national test facility for evaluating commercial load levelling batteries. The battery test modules have rated energies of 5–10 MWh and power ratings of 1 MW: the full-sized batteries are planned to have an available energy of 100 MWh energy and a power of 20 MW. The largest battery to date is a 40 MWh system which was installed in California in 1988. This is a lead–acid series–parallel array which operates at 2000 V and can supply the grid with 5000 A for up to 4 hours, thus achieving a power output of 10 MW. A 100 MWh lead–acid battery would occupy a building two and a half storeys high and an area of about 0.25 km^2 (Fig. 1.7). Of the advanced load levelling batteries, the ambient zinc–chlorine hydrate and the high temperature sodium sulphur systems are at the most advanced stage of development. Other high temperature systems such as sodium–antimony trichloride and lithium–iron sulphide are also being studied for load levelling applications. Another system under consideration is the redox flow cell: engineering and costing studies have been made for 10–100 MWh units.

1.5   Conclusions

The aims of the battery manufacturer have remained much the same since the beginnings of commercial exploitation. For small primary systems, the main goal is to provide higher energy and power density coupled with long shelf life and low cost. However, primary batteries are always an expensive

---

* In Scotland, the cost of domestic mains electricity is £0.0713/kWh (in 1997). A D-size Leclanché cell, delivering say 5 Wh, currently retails at £0.50. Thus, energy from the primary battery costs £100/kWh – a factor of over 1000 more expensive. For a 150 mWh zinc–silver oxide button cell, retailing at £1.50, the cost of energy is over £10 000/kWh!

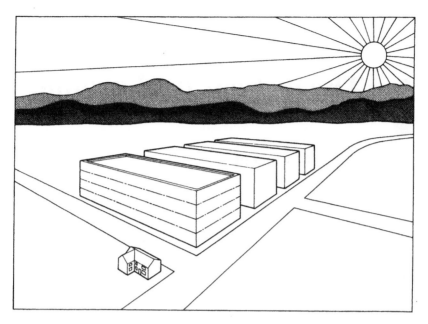

Fig. 1.7 Artist's impression of a 100 MWh load levelling battery. (By courtesy of I.D.C. Cargill, University of St Andrews.)

way of purchasing energy,* and further moves to replace many primary batteries with compatible rechargeable units are to be expected. For application in electric vehicles, energy density (again coupled with low cost) is the prime requirement. However, for energy storage, cost alone must be the overriding consideration. The overall cost of energy storage can be divided into two components. First, the initial capital cost of the battery must be considered, which includes raw materials, a contribution to research and development, investment in plant, administrative overheads and manufacturing costs. Provided that the battery has a lengthy cycle life, a high capital cost may be acceptable: a system capable of 1500 cycles and operated at 1 cycle/day, five times per week could be amortized over a 5–6 year period. The second component is determined by the efficiency of the charge/discharge cycle, i.e. by the ratio of energy supplied by the battery during discharge to the energy required to charge it. The extensive exploitation of batteries in a load levelling role must await the development of a new, long life, highly efficient system.

2 Theoretical background

Colin A Vincent

2.1 Introduction

Many chemical substances are able to exist in more than one oxidation state: that is to say, they can donate or accept electrons from other species. A net transfer of electrons such as

$$M^{x+}(aq) + P^{y+}(aq) \rightarrow M^{(x+1)+}(aq) + P^{(y-1)+}(aq) \qquad (2.1)$$

may occur spontaneously whenever the free energy of the whole system is reduced by such a process. An exchange of electrons can take place in homogeneous solution as above or, under suitable circumstances, electrons can be transferred to or from an electronic conductor. An electron transfer between an electronic conductor and some species in an electrolytic phase is known as an electrochemical reaction or electrode process. For such a transfer to proceed on a continuous basis, the principle of electroneutrality requires it to be associated with a second electrode process in which electron transfer takes place in the opposite direction. If the two compartments in the vessel shown in Fig. 2.1 contain solutions of $M^{x+}(aq)$ and $P^{y+}(aq)$, respectively, then connection of the two inert metal conductors through a resistor will permit charge flow round the circuit and hence formation of $M^{(x+1)+}(aq)$ and $P^{(y-1)+}(aq)$. As charge transfer (in the appropriate direction) at the interfaces produces exactly the same result as the homogeneous electron exchange, the whole cell process, including the electron flow through the load resistor, will also be spontaneous, driven by the net free energy change associated with the cell reaction. This is the essence of the electrochemical or galvanic cell, which is thus seen to be a device which allows the direct conversion of chemical into electrical energy.

The quantity of a chemical species reduced or oxidized in an electrochemical cell is related by Faraday's Laws to the total electric charge

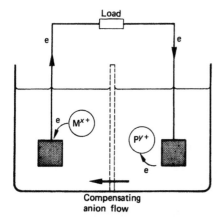

Fig. 2.1 Basic electrochemical cell. Interconnection of the two electrodes through the external load resistor allows the spontaneous cell reaction to proceed

transferred acoss the metal–solution interfaces. A current of i amps flowing in the circuit for a time of δt seconds is equivalent to the transfer of $i.\delta t$ coulombs of charge across any interface in the cell. Now if the oxidation of each M^{x+}(aq) ion involves the transfer of one electron from the solution phase to the electrode (Fig. 2.1), then the passage of $i.\delta t$ coulombs must correspond to the oxidation of

$$i\,\delta t/(Le_{o})$$

moles of M^{x+}(aq), where e_{o} is the charge on an electron (1.601×10^{-19}C), and L is the Avogadro constant, or number of atoms in a mole (6.022×10^{23}). The term (Le_{o}) is given the name Faraday's constant, F, and has a value of 96 490 C mol^{-1}. For the more general electrode process

$$M^{x+}(aq) - ne \rightarrow M^{(x+n)+}(aq)$$

each M^{x+}(aq) ion gives up n electrons to the external circuit in order to form the oxidized product. Here the number of moles of reactant M^{x+}(aq) consumed by the passage of a current, i, flowing for a period δt is given by

$$N_{M} = \frac{i\,\delta t}{nF} \tag{2.2}$$

Since the current passing through a cell is often a function of time, a more general form of equation (2.2) is

$$N_{M} = \frac{1}{nF} \int_{0}^{t} i\,\mathrm{d}t \tag{2.2'}$$

where the integration is carried out over the interval during which current passes. Conversely, if N_{M} represents the total number of moles of M^{x+}(aq) present, then the maximum capacity of the cell, i.e. the total amount of charge that could be supplied to the external circuit, is given by

$$Q_T = \int_0^t i \, dt = nF.N_M \tag{2.3}$$

Finally, recalling that current can only flow in a cell if there are two electrodes, the transfer of $i.\delta t$ coulombs in the appropriate direction at the second electrode (Fig. 2.1) brings about the reduction of $P^{y+}(aq)$ ions. If this second electrode process is

$$P^y + (aq) + n'e \rightarrow P^{(y-n')+}(aq)$$

then N_P, the number of moles of $P^{y+}(aq)$ reduced, is given by

$$N_P = \frac{i \, \delta t}{n'F} \tag{2.4}$$

Note that the maximum capacity of a cell which contains N_M moles of $M^{x+}(aq)$ and N_P moles of $P^{y+}(aq)$ is *the lesser* of the two quantities nFN_M and $n'FN_P$.

If

$$nN_M = n'N_P$$

the composition of the cell is exactly balanced, and the reactants at each electrode theoretically become exhausted at the same time. In contrast, if

$$nN_M > n'N_P$$

the cell is said to be positive-limited or cathode-limited, i.e. the maximum charge that can be supplied to the external circuit is determined by only one of the two active cell components, here $P^{y+}(aq)$. For operational reasons, most practical cells have such an imbalance in their make-up; this may be to prevent hazardous reactions such as the liberation of gases, or catastrophic failure caused by the puncture of a consumable container, at the end of the cell discharge.

2.2   The electrical double layer and the formation of electric potentials at interfaces

Whenever two phases which have significant electrical conductivities come into contact, there is generally some redistribution of charge – by charge transfer across the interface as described above, by adsorption of charged particles at the interface, by orientation of dipoles near the interface, etc. The region containing this distribution of charge is known as the electrical double layer, since the total charge excess on the one side of the interface must, for reasons of overall electroneutrality, be exactly equal and opposite to that on the other. When a metal is immersed in a solution of an electrolyte, the charge on the metal side of the interface corresponds to a surplus or deficit of electrons in the surface atomic layer. In most models, the metal is taken to have a perfect planar surface. In practice there are always grain boundaries and point defects present. Further, even in the case of a 'perfect' single crystal facet, 'reconstruction' low coordination surfaces may occur in order to reduce the surface energy. Such reconstruction may be reversed or altered if adsorption of electrolyte components occurs. The structure of the solution side of the double layer is even more complicated; it can be considered to consist of two regions (Fig. 2.2a–c):

• An inner layer which may contain partially desolvated anions which have strong chemical bonding interactions with the metal, together with oriented solvent molecular dipoles.

• A 'time averaged' diffuse layer where ions are attracted to or repelled from the interface by electrostatic forces, but which are also affected by thermal collisions. The effective depth of the diffuse layer is a function of electrolyte concentration, temperature, dielectric constant of the solvent, etc.

The formation of an electrical double layer at a metal–solution interface brings about a particular arrangement of atoms, ions and molecules in the region near the electrode surface, and an associated variation in electrical potential with distance from the interface. The double layer structure may significantly affect the rates of electrochemical reactions.

In the case of electrochemical cells capable of passing significant direct currents, the principal mechanism for the formation of such potential differences at phase boundaries within the cell is charge transfer, e.g. electron transfer between two metals or semiconductors, ion transfer

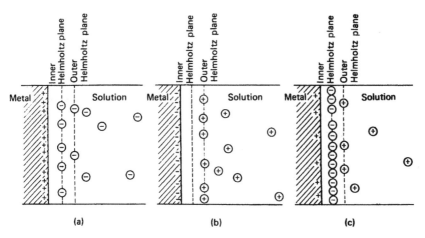

Fig. 2.2 Structure of the electric double layer under different conditions of electrode polarization: (a) metal positively charged, anions present at the inner Helmholtz plane (chemically interacting with metal) and in the diffuse double layer beyond the outer Helmholtz plane; (b) metal negatively charged, inner Helmholtz plane empty, cations in diffuse layer; (c) metal positively charged, strong adsorption of anions in inner Helmholtz plane, balancing cations in the diffuse layer

between a metal and a solution of its ions, etc. Consider a chemically inert metal, like platinum, immersed in a solution containing the oxidized and reduced forms of a chemical species, e.g. Fe^{3+} and Fe^{2+} ions. At the instant that the metal and solution are brought into contact, the two phases are uncharged and there is no electric potential difference between them. However, whenever they are brought into contact, charge transfer processes begin: some electrons from the conduction band of the metal are accepted by Fe^{3+} ions in the solution while some of the Fe^{2+} ions donate electrons from the t_{2g} orbital to the metal. If the rates of these two processes are unequal, the metal and solution phases become progressively charged: for example, if the Fe^{3+} ions are more readily reduced than are the Fe^{2+} ions oxidized, the metal soon gains a net positive charge (through losing more electrons than it gains) and the solution an equal negative charge (made up of the excess balancing anions left over, when Fe^{3+} ions are converted to Fe^{2+} ions). As the electric charge on the two phases builds up, so an electrical potential difference develops between them. However, the existence of such an electric potential difference affects the charge transfer rates. If the electrode (i.e. the metallic phase) acquires a relative positive charge, then the rate of reduction of Fe^{3+} ions will be depressed. At the same time, the rate of oxidation of Fe^{2+} ions will be increased. Thus, as the charge transfer processes proceed, the electric potential difference between the phases continues to alter until the rates of the processes

$$Fe^{2+} \xrightarrow{-e} Fe^{3+} \text{ and } Fe^{3+} \xrightarrow{+e} Fe^{2+}$$

become equal. At this point an equilibrium has been established and the electric potential difference between the phases is known as the equilibrium potential difference. Note that the charge transfer processes have not stopped. Rather, there is now an equal transfer of electrons to and from the metal so that there is no further *net* charge transfer between the phases, and an equilibrium double layer structure is established.

Two points may be made at this stage. First, the quantity of charge transferred between phases in order to establish an equilibrium potential difference is normally so small that the actual change in composition of the solution is negligible. For example, one can show that when a 1 cm^2 platinum electrode is immersed in a Fe^{2+}/Fe^{3+} solution, a net reduction of between 10^{-9} and 10^{-10} moles of Fe^{3+} takes place. Second, and as will be stressed later, the kinetics of the charge transfer process are very important, since if rates are slow, it may not be possible for a true equilibrium to be established.

An essentially similar description can be given for the build-up of a potential difference between a metal such as zinc and a solution containing Zn^{2+} ions, where the metal acts as the reduced form. Indeed, a similar charge redistribution or double layer formation occurs whenever two different electric conductors are placed in contact. For instance, when two different metals are connected together there will be a net flow of electrons from one to the other, resulting in a separation of charge between the two

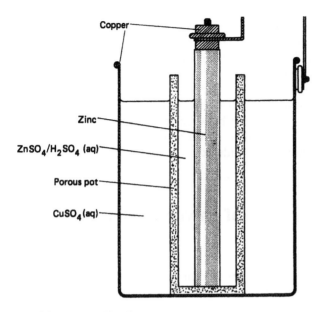

Fig. 2.3 The Daniell cell

metallic phases and consequently what is known as a contact potential difference is set up.

The historically important Daniell cell consists of a copper vessel containing a saturated solution of copper sulphate which forms one of the two electrode systems, and a central amalgamated zinc rod immersed in a zinc sulphate/sulphuric acid solution, which forms the other (Fig. 2.3). Gross mixing of the two solutions is prevented by a membrane or porous pot. It is assumed here that both electrodes are provided with copper terminals.

In this cell there are three significant interfacial potential differences (neglecting any small liquid junction potential difference):

Cu	Zn	Solution	Cu
(1)	(2)	(3)	

The Galvani potential, ϕ, of a phase defines the amount of electrical energy, $e\phi$, required to transport a charge e from an infinitely distant point in a vacuum to a hypothetical point in the interior of the phase where the charge would experience no 'chemical' forces exerted on it. Thus the Daniell cell voltage can be written as*

$$E_{cell} = {}^{Cu}\Delta^{Zn}\phi + {}^{Zn}\Delta^{Solution}\phi + {}^{solution}\Delta^{Cu}\phi \qquad (2.5)$$

---

* Note that ${}^{Cu}\Delta^{Zn}\phi$ is by no means negligible: from work function measurements it can be shown to have a value of about 0.25V.

Unfortunately, it is impossible to measure either the Galvani potential of a single phase or the Galvani potential difference between two phases of different composition. The only Galvani potential difference that can be measured in this cell is that existing between the two copper terminals. Thus it is not particularly instructive to develop further the concept of individual electric potential differences at phase boundaries within a galvanic cell. Instead, the relationship between the measurable cell voltage and the properties of the chemical processes associated with the running of the complete cell will now be considered.

2.3   Thermodynamics of galvanic cells

Energetics

Discharge of the Daniell cell causes zinc metal to ionize at one electrode and copper ions to deposit at the other. The net result of charge flow round the circuit is therefore equivalent to the reaction

$$Zn(s) + Cu^{2+}(aq) \rightarrow Zn^{2+}(aq) + Cu(s) \tag{2.6}$$

This is known as the cell reaction.

A cell is said to act reversibly if the net cell reaction is reversed when the current through the cell is made to flow in the opposite direction. When no current is being drawn, such a cell is in a true equilibrium state. Note that the absence of net current flow does not necessarily signify that a cell is in equilibrium. If an iron wire is placed in a solution of low pH, the most likely electron transfer reactions at the metal/solution interface are

$$2H^+(aq) + 2e \rightarrow H_2(g) \tag{2.7}$$

and

$$Fe(s) \rightarrow Fe^{2+}(aq) + 2e \tag{2.8}$$

At a particular potential, no net current at the interface will be observed. However, a cell containing such an electrode would obviously fail to fulfil the reversibility condition.

When no current is being drawn *from a reversible cell*, the potential difference across its terminals, or open circuit voltage (OCV), is known as the electromotive force or emf of the cell. The emf of any particular reversible cell is a quantitative measure of the tendency of the cell reaction to occur and may be related to the free energy change for this process. Let the emf of a reversible cell be balanced by a Poggendorff potentiometer (Fig. 2.4). At the balance point, where no current is flowing in the cell, the voltage E applied by the potentiometer is just sufficient to stop the cell reaction. (If it were increased any further, the normal cell discharge reaction would be reversed.) Let the applied voltage now be reduced to $(E - \delta)$.

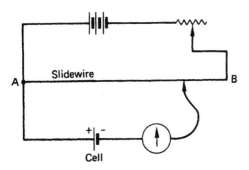

Fig. 2.4 The Poggendorff potentiometer: a circuit which permits the emf of a cell to be backed off so that the spontaneous cell reaction can be allowed to proceed, to be exactly halted (galvanometer at zero) or to be reversed

Under these circumstances, current flows and the work done (in joules) by the cell as one mole of reactants are converted to products would be equal to the product of the charge driven through the applied voltage (say nF coulombs) and the value of this potential difference ($E-\delta$ volts)

$$w = nF(E-\delta) \qquad (2.9)$$

where n is the number of moles of electrons transferred in one mole of reaction. Obviously the value of δ can be made as small as desired, so that the maximum work that may be obtained from the cell for one mole of reaction is

$$w_{max} = nFE \qquad (2.10)$$

By definition, $w_{max} = -\Delta G$, where ΔG is the free energy change associated with one mole of reaction and hence

$$\Delta G = -nFE \qquad (2.11)$$

In practice, passage of any finite amount of electricity results in a certain degree of irreversibility as some of the electrical energy is dissipated as joule heat in the internal resistance of the cell, or at the electrodes where application of a potential difference is required to drive the current at the desired rate. Galvanic cells can only supply electric work equal to the free energy change of the cell reaction when the current flowing tends to zero. The cell voltage under load is always smaller than the open circuit voltage (OCV), so that only part of the thermodynamically available work can be utilized. The various causes of internal resistance or 'polarization' of the cells are considered later.

The enthalpy change associated with a cell reaction is a state function and hence is independent of whether the reaction is being carried out reversibly or not. If an exothermic reaction is carried out completely irreversibly, the total enthalpy change ΔH will appear as heat given out to the surroundings:

$$\text{Heat given out} = q_{out} = -\Delta H \qquad (2.12)$$

On the other hand, if some work is done by the cell (e.g. by turning an electric motor in order to lift a weight) then

$$w_{out} + q_{out} = -\Delta H \qquad (2.13)$$

The second law of thermodynamics determines how much work can be extracted from a process: for maximum work output there is an associated minimum heat output:

$$(w_{out})_{max} + (q_{out})_{min} = -\Delta H \qquad (2.14)$$

The term $(q_{out})_{min}$ may be identified with $-T\Delta S$, where ΔS is the entropy change of the cell reaction. If ΔS is positive, the cell will cool down as it operates, or take in heat from the surroundings. If the same cell operates irreversibly, it will cool down less, or take in less heat from the surroundings.

Schematic representation of a galvanic cell

A galvanic cell may be represented by writing the composition of the individual phases of which it is composed in the order in which they are connected. A convenient notation is to indicate each phase boundary by a vertical line. The zone of contact of two solution phases is indicated by a double vertical line. (This region is usually associated with a small non-equilibrium potential difference caused by charge separation resulting from the unequal rates of diffusion of ions across it.) The accepted convention is for a cell diagram to be drawn so that the cell reaction under consideration, whether spontaneous or not, takes place when positive charge flows through the cell from left to right. The cell voltage is then equal in sign and magnitude to the voltage of the right-hand terminal, taking that of the left as zero. E_{cell}, the cell emf, is positive when the cell reaction proceeds spontaneously on connecting the two terminals together. Thus if m_1 and m_2 refer to the molal concentrations of $Zn^{2+}(aq)$ and $Cu^{2+}(aq)$, respectively, it is equally correct to write for the Daniell cell either

$$Zn(s)|Zn^{2+}(aq),m_1||Cu^{2+}(aq),m_2|Cu(s) \qquad E_{cell} = +1.1 \text{ V}$$

which describes the spontaneous discharge process

$$Zn(s) + Cu^{2+}(aq),m_2 \rightarrow Zn^{2+}(aq),m_1 + Cu(s)$$

or

$$Cu(s)|Cu^{2+}(aq),m_2||Zn^{2+}(aq),m_1|Zn(s) \qquad E_{cell} = -1.1 \text{ V}$$

which describes the non-spontaneous charging reaction

$$Zn^{2+}(aq),m_1 + Cu(s) \rightarrow Zn(s) + Cu^{2+}(aq),m_2$$

In practice, the first of these alternatives is generally used. The implication of this is that galvanic cells are written with the positive electrode on the

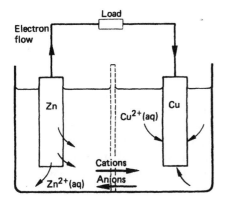

Fig. 2.5 Ion and electron flow in a schematic Daniell cell

right, electron flow in the external circuit is from left to right, and in the solution phase, cation flow is from left to right and anion flow from right to left (Fig. 2.5).

It is convenient at this point to introduce the terms 'anode' and 'cathode'. A cathode is defined as the electrode to which electrons flow from the external circuit – i.e. as the electrode at which reduction takes place. Conversely, electrons flow from the anode to the external circuit: oxidation takes place at the anode. The point of difficulty rests in the fact that in a galvanic cell under spontaneous discharge, the cathode is the positive electrode, whereas in an electrolytic cell or galvanic cell under charge, it is the negative electrode (Fig. 2.6). Any possible confusion can be avoided if the electrode which acts as anode during the discharge of a secondary cell is always referred to as the negative plate, or negative, while the electrode which acts as cathode during discharge is called the positive plate, or positive.

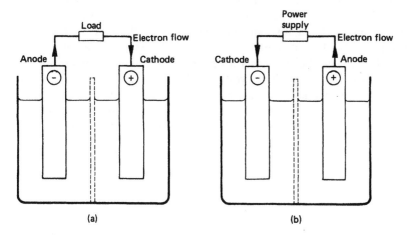

Fig. 2.6 Anode and cathode in a cell during (a) spontaneous discharge and (b) charge

Half-reactions, half-cells and electrode potentials

The emf of a reversible cell can be regarded either as a function of the free energy change associated with the overall cell reaction or as a sum of the Galvani potential differences between phases within the cell. It was noted above, however, that individual Galvani potential differences between non-identical phases could not be measured and it is therefore impossible to resolve a cell emf into its interphasial components. On the other hand, every cell reaction consists of an oxidation and a reduction process and thus can be considered as the sum of two notional 'half-reactions' occurring in notional 'half-cells'. For example, the Daniell cell can be visualized as consisting of the half-cells

$$Cu(s)|Zn(s)|Zn^{2+}(aq),m_1$$

and

$$Cu(s)|Cu^{2+}(aq),m_2$$

If cells are constructed by making different combinations of half-cells, then the cell emf values obey an additive law. It is therefore convenient to combine half-cells with a single reference half-cell and thus obtain a series of related emf values compared to the reference half-cell taken as zero.

The universally accepted primary reference half-cell is the standard hydrogen electrode. The electrode consists of a noble metal (platinized platinum) dipping into a solution of hydrogen ions at unit activity and saturated with hydrogen gas at 1 bar (i.e. 1×10^5 Pa, which in practical terms may be taken to be equal to 1 atmosphere). In practice such a standard electrode cannot be realized, but the scale it defines can.

The electrode potential is defined as the potential difference between the terminals of a cell constructed of the half-cell in question and a standard hydrogen electrode (or its equivalent) and assuming that the terminal of the latter is at zero volts. Note therefore that the electrode potential is an observable physical quantity and is unaffected by the conventions used for writing cells. The statement '. . . the electrode potential of zinc is -0.76 volts . . .' implies only that a voltmeter placed across the terminals of a cell consisting of standard hydrogen electrode and the zinc electrode would show this value of potential difference, with the zinc terminal negative with respect to that of the hydrogen electrode. An electrode potential is *never* a 'metal/solution potential difference', not even on some arbitrary scale.

An alternative way of presenting data on cells with a standard hydrogen electrode is in terms of the reduction potential or oxidation potential of the cell reaction. For the cell reaction

$$Zn^{2+}(aq) + H_2(g) \rightarrow Zn(s) + 2H^+(aq) \tag{2.15}$$

the notional half-reaction is written as

$$Zn^{2+}(aq) + 2e \rightarrow Zn(s) \tag{2.16}$$

and the emf of the corresponding cell

$$Pt(s),H_2(g),1 \text{ bar}|H^+(aq),a=1||Zn^{2+}(aq),a_{Zn^{2+}}|Zn(s)$$

is called the reduction potential, E_{Red}. If the cell reaction (and hence the half-reaction) is reversed, the sign of the half-cell potential must also be reversed, so that for

$$Zn(s) + 2H^+(aq) \rightarrow Zn^{2+}(aq) + H_2(g) \tag{2.17}$$

the notional half-reaction is written as

$$Zn(s) \rightarrow Zn^{2+}(aq) + 2e \tag{2.18}$$

and the emf of the corresponding cell

$$Zn(s)|Zn^{2+}(aq),a_{Zn^{2+}}||H^+(aq),a=1|H_2(g),1 \text{ bar, Pt(s)}$$

is defined as the oxidation potential, E_{Ox}. Thus

$$(E_{Ox})_{Zn} = - (E_{Red})_{Zn} \tag{2.19}$$

It is perhaps best to consider oxidation and reduction potentials as properties of a half-reaction rather than of a half-cell. (Some confusion might be avoided if tables of reduction potentials were written in terms of the complete cell reaction rather than as notional half-reactions.) It can be seen that the electrode potential has the same value as the reduction potential of the half-reaction. For the Daniell cell, the following relationships exist:

$$Zn(s)|Zn^{2+}(aq)||Cu^{2+}(aq)|Cu(s)$$

Emf of cell $= E_{cell}$

$\qquad\qquad$ = electrode potential of copper − electrode potential of zinc

$\qquad\qquad = (E_{Red})_{Cu} - (E_{Red})_{Zn}$

$\qquad\qquad = (E_{Red})_{Cu} + (E_{Ox})_{Zn}$

$\qquad\qquad = (E_{Ox})_{Zn} - (E_{Ox})_{Cu} \tag{2.20}$

Rules for the manipulation of reduction (or oxidation) potentials

1. A half-reaction can be multiplied through by any number without affecting the reduction (or oxidation) potential, e.g.

$$Fe^{3+}(aq) + e \rightarrow Fe^{2+}(aq); \quad E_{Red} = +0.771 \text{ V}$$
$$3Fe^{3+}(aq) + 3e \rightarrow 3Fe^{2+}(aq); \quad E_{Red} = +0.771 \text{ V}$$

2. Subtraction of reduction (or of oxidation) potentials of half-reactions having the same number of electrons gives the emf of the corresponding cell reaction, e.g.

$$Cl_2(g) + 2e \rightarrow 2Cl^-(aq); \quad E_{Red} = +1.360 \text{ V}$$
$$Zn^{2+}(aq) + 2e \rightarrow Zn(s); \quad E_{Red} = -0.763 \text{ V}$$

Therefore for

$$Cl_2(g) + Zn(s) \rightarrow 2Cl^-(aq) + Zn^{2+}(aq); \quad E_{cell} = +2.123 \text{ V}$$

3. Subtraction of reduction (or of oxidation) potentials of two half-reactions not having the same number of electrons does *not* give the reduction (or oxidation) potential of the resulting new half-reaction.

For

$$Cu^{2+}(aq) + e \rightarrow Cu^+(aq); \quad E_{Red} = +0.150 \text{ V} \qquad \text{(i)}$$

for

$$Cu^{2+}(aq) + 2e \rightarrow Cu(s); \quad E_{Red} = +0.336 \text{ V} \qquad \text{(ii)}$$

and for

$$Cu^+(aq) + e \rightarrow Cu(s); \quad E_{Red} = +0.522 \text{ V} \qquad \text{(iii)}$$

Subtracting half-reaction (i) from half-reaction (ii) gives half-reaction (iii); subtracting the reduction potentials gives +0.187 V, which is *not* the reduction potential of (iii).

However,

$$\Delta G_2 = -2 \times 0.336 \times F$$
$$\Delta G_1 = -1 \times 0.150 \times F$$
$$\Delta G_3 = \Delta G_2 - \Delta G_1 = -(2 \times 0.336 - 0.150) F$$
$$(E_{Red})_{(iii)} = 2 \times 0.336 - 0.150 = +0.522 \text{ V}$$

Dependence of emf on concentration

The dependence of emf on the activities or concentrations of the products and reactants of a cell reaction follows directly from a consideration of the relationship between emf and free energy change. For the cell reaction

$$v_A A + v_B B + \ldots \rightarrow v_P P + v_Q Q + \ldots \qquad (2.21)$$

if v_i is the stoichiometric number and μ_i is the chemical potential of a species involved, then

$$\Delta G = \sum_i v_i \mu_i = \sum_i v_i (\mu_i^\ominus + RT \ln a_i) \qquad (2.22)$$

Here $\mu_i^\ominus$ is the chemical potential of i in its standard state and a_i is its molal activity. This equation can be recast in the form known as the van't Hoff reaction isotherm:

$$\Delta G = \Delta G^\ominus + RT \ln \left[\frac{a_P^{v_P} a_Q^{v_Q} \cdots}{a_A^{v_A} a_B^{v_B} \cdots} \right] \qquad (2.23)$$

Hence the emf of a cell is given by

$$E = E^{\ominus} - \frac{RT}{nF} \ln \left[\frac{a_P{}^{v_P} a_Q{}^{v_Q} \dots}{a_A{}^{v_A} a_B{}^{v_B} \dots} \right] \qquad (2.24)$$

$E^{\ominus}$ is the standard emf of the cell and is the equilibrium voltage when all the cell components are in their standard states: solution species have unit molal activities, gases have pressures of 1 bar and solid phases are in their most stable form. Equation (2.24) is known as the Nernst equation for a galvanic cell. In an approximate form of this equation, molal activities are replaced by molar concentrations to give

$$E = E^{\ominus} - \frac{RT}{nF} \ln \left[\frac{[P]^{v_P} [Q]^{v_Q} \dots}{[A]^{v_A} [B]^{v_B} \dots} \right] \qquad (2.24')$$

(The values of the standard emf in eqs 2.24 and 2.24′ are slightly different.) Since a half-reaction represents a cell involving a standard hydrogen electrode, the Nernst equation may also be applied to electrode potentials. Thus for the half-reaction

$$Zn^{2+}(aq), a_{Zn^{2+}} + 2e \rightarrow Zn(s)$$

the electrode potential is given by

$$E = E^{\ominus}_{Zn} - \frac{RT}{2F} \ln \frac{a_{Zn(s)}}{a_{Zn^{2+}}}$$

$$= E^{\ominus}_{Zn} - \frac{RT}{2F} \ln \frac{1}{a_{Zn}}$$

$$= E^{\ominus}_{Zn} + \frac{RT}{2F} \ln a_{Zn^{2+}}$$

$$\approx E^{\ominus}_{Zn} + 0.0128 \ln [Zn^{2+}] \text{ at } 25°C \qquad (2.25)$$

Note that $a_{Zn(s)} = 1$ since zinc is a pure substance. A more complex relationship is required if one component is a non-stoichiometric solid which can exist over a range of compositions. $E^{\ominus}_{Zn}$ is known as the standard electrode potential of zinc.

The dependence of cell emf on the concentrations of reactants and products of the cell reaction is of fundamental importance in the understanding and design of certain practical battery systems. As a cell undergoes discharge, reactants are steadily converted to products until one of the reactants species is completely exhausted. Consider the cell reaction A + B → C + D, where A–D are all solution species. For this system,

$$E_{cell} = E^{\ominus}_{cell} - \frac{RT}{nF} \ln \left(\frac{a_C \, a_D}{a_A \, a_B} \right)$$

At low discharge levels the concentration of reactants will be large and that of the products small, so that E_{cell} will be relatively high. As the discharge continues, the thermodynamic-controlled open circuit voltage (i.e. the cell emf) will rapidly fall. If a battery is to be designed to have only a small decrease in open circuit voltage from the initial open circuit value across the entire discharge, it is best to select reactants and products whose activities remain effectively constant. This is the situation when these components are pure materials or the major constituent in a phase whose composition does not vary significantly. In Figs 2.7 and 2.8 the open circuit voltage is plotted as a function of percentage discharge for the cells

$$Zn(s)|ZnO(s)|KOH(aq)|HgO(s)|Hg(l)$$

for which the cell reaction is

$$Zn(s) + HgO(s) \rightarrow ZnO(s) + Hg(l)$$

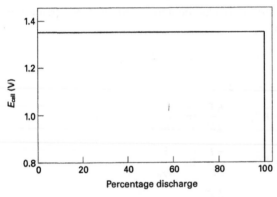

Fig. 2.7 Open circuit voltage as a function of percentage discharge for the cell Zn(s)|ZnO(s)|KOH(aq)|HgO(s)|Hg(l)

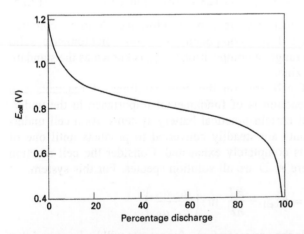

Fig. 2.8 Open circuit voltage as a function of percentage discharge for the cell Pt(s)|Fe^{2+}(aq),Fe^{3+}(aq),||Ce^{3+}(aq),Ce^{4+}(aq),|Pt(s), where the concentration of each of the four solution species is 0.1 mol/dm^3

and

$$Pt(s)|Fe^{2+}(aq),Fe^{3+}(aq)||Ce^{3+}(aq),Ce^{4+}(aq)|Pt(s)$$

where the concentration of each of the four solution species is 0.1 mol/dm^3 and for which the cell reaction is

$$Fe^{2+}(aq) + Ce^{4+}(aq) \rightarrow Fe^{3+}(aq) + Ce^{3+}(aq)$$

Dependence of emf on temperature and pressure

From the identity

$$dG = - SdT + Vdp$$

it is seen that

$$\left(\frac{\partial \Delta G}{\partial T}\right)_p = -\Delta S \quad \text{or} \quad \left(\frac{\partial E}{\partial T}\right)_p = \frac{\Delta S}{nF} \tag{2.26}$$

$$\left(\frac{\partial \Delta G}{\partial p}\right)_T = -\Delta V \quad \text{or} \quad \left(\frac{\partial E}{\partial p}\right)_T = \frac{-\Delta V}{nF} \tag{2.27}$$

Cell emf values may therefore have positive or negative temperature coefficients, depending on the sign of the entropy change associated with the cell reaction. It should be noted that unless gases are involved in the latter, the entropy change and hence the temperature coefficient of the emf is usually very small. A useful practical form for the temperature dependence is

$$E_T = E_{298} + \frac{(T - 298)\Delta S}{nF} \tag{2.28}$$

where T is the Celsius temperature.*

Again, for cell reactions involving only condensed phases, the emf can be regarded as virtually independent of applied pressure. However, when there is a change in the number of moles of gas during a cell reaction the volume change, ΔV, cannot be neglected:

$$E_p = E_{p^\ominus} - \frac{1}{nF} \int_{p^\ominus}^{p} \Delta V dp \tag{2.29}$$

where $p^\ominus$ represents the standard pressure. For a cell such as

$$Pt(s),H_2(g),p|HCl(aq),m|AgCl(s)|Ag(s)$$

for which the cell reaction is

---

* It is well known that the *working* voltage of a practical cell under load usually rises with an increase in temperature. However, this is almost entirely due to a reduction in the internal resistance of the cell caused by an increase in the conductivity of the electrolytic phase and in the diffusion rates of the electroactive species.

$$\tfrac{1}{2}H_2(g) + AgCl(s) \rightarrow H^+(aq) + Ag(s) + Cl^-(aq) \qquad (2.30)$$
$$\Delta V = -\tfrac{1}{2}\bar{V}_{H_2}$$

where $\bar{V}_{H_2}$ is the molar volume of hydrogen. For pressures up to some tens of atmospheres it can be assumed that the ideal gas law holds, so that

$$\Delta V = -\frac{RT}{2p}$$

Hence

$$E_p = E_{p^{\ominus}} + \frac{RT}{2nF} \int_{p^{\ominus}}^{p} \frac{dp}{p}$$

$$= E_{p^{\ominus}} + \frac{RT}{2nF} \ln p \qquad (2.31)$$

At higher pressures more accurate equations of state are required and ΔV generally becomes a more complex function of the pressure.

Electrode types

Electrodes or half-cells are often classified into three main types. The classification is mainly historical since there are no fundamental differences between them.

Electrodes of the first kind

Electrodes of this form have a minimum of two phases, namely a metal (either in pure form, or in an amalgam or solid solution) in contact with an electrolyte phase (solution, melt, solid electrolyte, glass, etc.) containing cations of the metal.

For the half-reaction

$$M^{n+}, a_{M^{n+}} + ne \rightarrow M, a_M \qquad (2.32)$$

the electrode potential is given by

$$E = E^{\ominus}_{M^{n+}/M} + \frac{RT}{nF} \ln \frac{a_{M^{n+}}}{a_M} \qquad (2.33)$$

If the phase containing M is pure metal then $a_M = 1$.

Electrodes of the second kind

In these systems there are a minimum of three phases: a metal, an adherent layer of sparingly soluble salt of the metal, and an electrolyte phase containing the same anion as the latter. Equilibrium exists both between the metal and cations of the sparingly soluble salt and between anions in the salt and those in solution. For the half-reaction

$$AgCl(s) + e \rightarrow Ag(s) + Cl^-(aq) \tag{2.34}$$

the electrode potential is given by

$$E = E^{\ominus}_{Ag^+/Ag} + \frac{RT}{F} \ln a_{Ag^+}$$

$$= (E^{\ominus}_{Ag^+/Ag} + \frac{RT}{F} \ln K_s) - \frac{RT}{F} \ln a_{Cl^-}$$

$$= E^{\ominus}_{Ag/AgCl/Cl^-} - \frac{RT}{F} \ln a_{Cl^-} \tag{2.35}$$

where K_s is the thermodynamic solubility product of AgCl.

The most common electrode of this type is the saturated calomel electrode (SCE) which consists of mercury in contact with a layer of insoluble Hg_2Cl_2 immersed in a saturated aqueous solution of KCl. The SCE is used as a *secondary standard* reference electrode. At 25°C it has an electrode potential of + 0.2415 V.

Redox electrodes

In redox electrodes an inert metal conductor acts as a source or sink for electrons. The components of the half-reaction are the two oxidation states of a constituent of the electrolytic phase. Examples of this type of system include the ferric/ferrous electrode where the active components are cations, the ferricyanide/ferrocyanide electrode where they are anionic complexes, the hydrogen electrode, the chlorine electrode, etc. In the gaseous electrodes equilibrium exists between electrons in the metal, ions in solution and dissolved gas molecules. For the half-reaction

$$Ox(aq) + ne \rightarrow Red(aq) \tag{2.36}$$

the electrode potential is given by

$$E = E^{\ominus}_{Ox/Red} + \frac{RT}{nF} \ln \frac{a_{Ox}}{a_{Red}} \tag{2.37}$$

Where one component is a gas, it is assumed that equilibrium exists between the gas phase and gas molecules dissolved in solution, so that the activity of the latter is given by the partial pressure in the gas phase. Thus for

$$\tfrac{1}{2}Cl_2(g), p_{Cl_2} + e \rightarrow Cl^-(aq), a_{Cl^-} \tag{2.38}$$

$$E = E^{\ominus}_{Cl_2/Cl^-} + \frac{RT}{F} \ln \frac{(p_{Cl_2})^{\frac{1}{2}}}{a_{Cl^-}} \tag{2.39}$$

Note that homogeneous electron transfer in the solution phase is negligible in the case of half-reactions involving gases, but that it may be very rapid for ionic redox systems.

Indirect values of electrode potentials

It must be realized that because of kinetic limitations, most half-cells that can be written cannot be the basis of a practical cell which will display the appropriate emf. It has however proved convenient to include such half-equations in tables of redox potentials if their emf could be evaluated in some other way. *In a large number of cases electrochemical data are not used at all.* Rather, partial molar heats and entropies of the species involved are determined by calorimetric methods and these are used to derive $\Delta G^\ominus$ for the cell reactions. $E^\ominus_{cell}$ values can then be calculated.

For example, consider how the standard potential for the process

$$2CO_2(g) + 2H^+(aq) + 2e \rightarrow H_2C_2O_4(aq)$$

might be found. This half-cell reaction implies the cell reaction

$$2CO_2(g) + H_2(g) \rightarrow H_2C_2O_4(aq)$$

If the free energies of formation of oxalic acid and carbon dioxide are known, then

$$\Delta G^\ominus = (\Delta G^\ominus_f)_{H_2C_2O_4} - 2(\Delta G^\ominus_f)_{CO_2} - 0$$

$$= (-695.0) - 2(-395.3)$$

$$= +95.68 \text{ kJ/mol}$$

and hence

$$E^\ominus_{cell} = \Delta G^\ominus/nF = -0.495 \text{ V}$$

Liquid junction potentials

When a cell contains a boundary between two electrolytic solutions of different composition or concentration, a liquid junction potential is developed due to the diffusion of the various components at characteristic rates in the boundary zone. Cells with liquid junction potentials cannot be regarded as reversible or equilibrium systems; rather, a constant liquid junction potential may be interpreted as indicating a steady state where an effectively time-independent charge separation has been developed. Diffusion, however, tends to be a slow process, so that the structure of the zone may hardly change. This is even more true when a separator or porous insulating matrix is used to immobilize the solution in the boundary layer, or where gelling agents are added to the electrolytes. Values of the liquid junction potential are difficult to calculate except for special cases, e.g. for the boundary between solutions having different concentrations of the same components. Liquid junction potentials are highest when there is a large concentration difference between the phases or where cation and anion mobilities are notably different. Under these circumstances liquid junction potentials of up to 50 mV may occur. However, in practical battery systems, any liquid junction potential is likely to be considerably lower and is almost always neglected.

2.4   Current flow in an electrochemical cell

Relationship between current and reaction rate

Within any cell undergoing charge or discharge, one can consider at least three forms of charge transmission:

- electron flow in the electronic conductors (e.g. electrode materials, current collectors, terminal connectors, load resistors, etc.);

- ion flow in the electrolyte (which may be an aqueous solution, solid electrolyte, molten salt, etc.);

- charge transfer reactions at the electrode/electrolyte interfaces (e.g. $Zn(s) \xrightarrow{-2e} Zn^{2+}(aq)$).

Since in the steady state, it is necessary to maintain a condition of electro-neutrality in any macroscopic part of the system, the total charge flux through all cross-sections of the circuit must be the same. In particular, the rate of electron flow in the external circuit is equal to the rate of charge transfer at each electrode/electrolyte interface.

For the general electrode process

$$Ox + ne \rightarrow Red$$

nF coulombs are required to reduce 1 mole of Ox, so that the rate of reduction of Ox at the electrode is given by

$$\left(\frac{\text{Rate of passage of electrons into electrode from external circuit}}{nF} \right)$$

i.e.

$$\text{Rate of reduction of Ox} = i/nF \text{ mol s}^{-1} \qquad (2.40)$$

At the other electrode, $Red' - n'e \rightarrow Ox'$, and again

$$\text{Rate of oxidation of Red}' = i/n'F \text{ mol s}^{-1} \qquad (2.40')$$

It should be noted that the rate of charge transfer at the electrodes (i.e. the rates of the electrochemical processes) may be given directly by the reading on an ammeter inserted in the external circuit. If the current is a function of time, it is still possible to apply the above idea of 'flux continuity' over a succession of small time intervals. It may happen that one of the various rate processes involved in charge transport in different components of a cell, is unable to maintain as high a rate as the others.

Under these circumstances it becomes the current limiting process for the cell.

The question now arises as to what factors are responsible for determining the rates at which the various cell processes occur. Thermodynamic arguments permit the feasibility of overall cell reactions to be predicted, but give no information on rates. To understand the latter it is necessary to consider the effects on various parts of the cell of forcing the cell voltage to assume a value different from that of the equilibrium emf. It has been shown above that in the Daniell cell at equilibrium, charge transfer across the zinc/solution interface can be described in terms of processes

$$Zn^{2+}(aq) \xrightarrow{2e} Zn(s)$$

and

$$Zn(s) \xrightarrow{-2e} Zn^{2+}(aq)$$

occurring at equal rates. The first of these may be considered as constituting a cathodic flow of charge across the interface, i.e. as a cathodic current, i_c. Similarly for the oxidation of metallic zinc to zinc ions, there is a charge flow which may be described as an anodic current, i_a. At equilibrium i_c is equal to i_a and this equilibrium current is known as the exchange current, i_o. The net current i, given as $(i_c - i_a)$ by convention, is of course zero at equilibrium. In the same way, at the copper electrode at equilibrium,

$$i_c' = i_a' = i_o'$$

(Note that it would be most improbable for i_o and i_o' to be equal.) Now let a constant voltage source be placed across the cell so that the zinc is forced to assume a less negative potential and the copper a less positive potential than their equilibrium values. The altered electric field across the zinc/solution interface makes it easier for zinc atoms to be oxidized but hinders the reduction of zinc ions. Hence i_a increases from its equilibrium value while i_c decreases: therefore

$$|i_{Zn}| = i_c - i_a \neq 0$$

At the copper electrode, reduction of Cu^{2+} is favoured and oxidation of Cu atoms is restricted, so that net cathodic flow occurs. Finally, to prevent a build up of Zn^{2+} ions near the zinc/electrolyte interface and of SO_4^{2-} counter ions near the copper, a flux of ions must take place in the electrolytic phase to balance the charge transfer processes at the interfaces. To maintain the flux continuity condition, the applied voltage difference becomes distributed in such a way that*

$$|i_{Zn}| = |i_{Cu}| = |i_{ion\ flow}| \tag{2.41}$$

* The voltage becomes distributed just as in the case of a potential, E, applied to three resistors in series:

$i = E/(R_1 + R_2 + R_3)$ and $E_1 = ER_1/(R_1 + R_2 + R_3)$, etc.

$R_1 \quad R_2 \quad R_3$

For spontaneous discharge, the overall cell voltage must be reduced from its equilibrium value, as would happen if a load resistor were connected to the terminals. If a potential difference greater than the emf were applied (i.e. one making the cathode more positive and the anode more negative), the net result would be a current flow in the reverse direction, causing a net charging of the cell.

Polarization losses

Knowledge of the amount by which the voltage of a cell, delivering a particular level of current, deviates from its equilibrium value is of central importance in assessing the performance of a practical battery system. This polarization voltage, E_{pol}, can be associated with two principal causes:

- 'ohmic' or 'iR' drop in the bulk of the electrolyte phases, separators and sometimes in the electrode phases and connectors;

- 'electrode losses' which include the 'activation overvoltage', connected with the charge transfer step and/or nucleation and crystallization processes at each electrode/electrolyte interface, and the 'concentration overvoltage' related to the depletion or accumulation of electroactive material near the electrode surfaces.

In practical batteries, especially those employing porous phases, it is not always possible to separate ohmic and electrode losses clearly.

Note that for cell discharge,

$$E_{cell} = E_{thermodynamic} - E_{pol} \tag{2.42}$$

while for a cell undergoing charge,

$$E_{cell} = E_{thermodynamic} + E'_{pol} \tag{2.43}$$

Ohmic potential drop

The source of ohmic potential drop is the internal resistance of the bulk phases within the cell. If the current distribution is uniform, then for a phase with conductance σ (S m^{-1}), the resistance is $R = x/A\sigma$, where x is the thickness of the phase and A is its cross-sectional area. Thus for the passage of a current i through a cell with j sequential phases,

$$E_{ohmic} = i \sum_j (x_j/A_j \sigma_j) \tag{2.44}$$

In practice, current distribution rarely approaches a uniform distribution. Instead it is a complex function of cell geometry, characteristics of the electrode surfaces, etc. It is sometimes possible to determine experimentally an effective or average value of R_{ohmic} using transient perturbation techniques. However, in many situations it is difficult to distinguish ohmic and electrode polarization, for instance where the electroactive material is

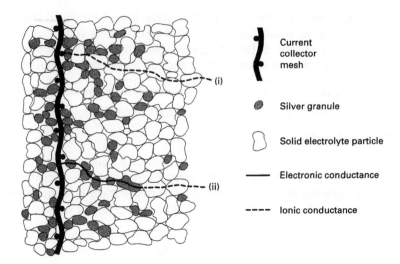

Current
collector
mesh

⬤ Silver granule

Solid electrolyte particle

—— Electronic conductance

- - - - Ionic conductance

Fig. 2.9 Current pathways in a solid state cell; the electrode consists of granular silver metal and solid electrolyte

also responsible for carrying all the current in the electrolyte phase, in the situation where porous electrodes are involved, or where electrode and electrolyte phases are finely mixed, as in some solid-state cells. In Fig. 2.9 two different conduction pathways are shown in a cell where the electrode material is granular silver ground together with a silver ion conducting solid electrolyte powder, and compressed to minimize electrode polarization by forming a large interfacial contact area. A very similar situation occurs with a porous electronically conducting electrode substrate and a liquid solution electrolyte.

It is common in many practical battery designs to 'immobilize' a liquid electrolyte phase within a porous solid insulator. The electrolyte conductivity and ohmic loss in such a system are determined by the number of pores, their size, shape and tortuosity. The tortuosity coefficient, β, is defined as the ratio of the mean distance covered by an ion traversing a porous matrix, to the direct distance of one side of the matrix to the other. The relative reduction in the conductivity of an electrolyte solution caused by confining it in a porous solid is called the conductivity attenuation, Θ. For a matrix of uniform cylindrical pores it is given by

$$\Theta = \pi N_p \, r_p^2 / \beta \qquad (2.45)$$

where N_p is the number of pores (with radius r_p) per unit area of the solid matrix, and

$$\kappa_M = \Theta \kappa \qquad (2.46)$$

where κ is the conductance of the free electrolyte solution and κ_M is its conductance when immobilized.

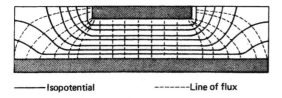

——— Isopotential　　　------Line of flux

Fig. 2.10 Potential distribution in a cylindrical cell with electrodes of unequal area

In most modern practical batteries, a major part of polarization loss at moderately high current densities is due to ohmic potential drop. Considerable attention is therefore given during the design of a battery to:

- obtaining maximum ionic conductance in the electrolyte phases (a particular problem in certain non-aqueous and solid electrolyte cells);

- reducing the resistivities of electrode materials (especially when they are in extended form, such as sintered powders, etc.) – for example by adding finely divided conductive carbon such as acetylene black – and supplying adequate current collectors, inter-cell connectors, etc.;

- balancing the advantages from an electrode polarization viewpoint of high interfacial areas against the drawbacks of high resistance paths in the bulk phases;

- optimizing cell geometry so as to minimize the effective distance between electrodes, while maintaining uniform current distribution.

This last consideration, namely uniform current distribution, is frequently of critical importance since variable current densities and the associated non-uniform potential distribution can result in localized depletion of reactants, shape changes in electrodes, side reactions, low electrode material utilization, etc. A simple example is shown in Fig. 2.10 where the potential distribution in a cylindrical cell with unequal electrode areas is shown. Very high current densities may occur near the circumference of the smaller electrode. As has been pointed out above, current flow in a cell is determined by ohmic loss in the electrolyte and electrode polarization losses at the electrode/electrolyte interfaces. If the latter is high, then the effect of uneven ohmic potential gradients becomes unimportant. However, under normal circumstances, ohmic loss is substantial and great care must therefore be taken during the design stage to try to establish uniform current densities at the electrode surfaces. In practice, the current density is affected by both potential and concentration gradients, so that theoretical treatments have been confined to rather simple models.

Electrode kinetics and electrode polarization losses

Polarization losses associated with an electrode process are termed 'overvoltage'. The overvoltage η is defined as

$$\eta = E - E_{\text{eq}} \tag{2.47}$$

where E_{eq} is the equilibrium (zero current) potential of an electrode and E is its potential when a certain current is being passed. Since an overall electrode process can be subdivided into a number of consecutive steps, involving transport of the electroactive species from the bulk of the electrolyte phase to the electrode surface, interfacial charge transfer, etc., a number of different effects may contribute to the total overvoltage. Of fundamental importance is the interfacial charge transfer step whose rate is directly controlled by the potential difference across the double layer and hence by the electrode potential. 'Charge transfer' is considered here to include the whole process of chemical bond formation and scission, solvation changes, etc. which must accompany the successful transfer of an electron or ion across an electrode/solution interface. In addition to the basic interfacial charge transfer, many electrode processes involve a further step in the vicinity of the phase boundary, e.g. electrocrystallization, in which atoms produced in the charge transfer step are incorporated into the lattice, formation of a new phase, or some slow chemical reaction, whose rate is independent of the electrode potential. Finally all electrode processes must involve mass transport whereby materials consumed or formed in the charge transfer step are moved to or from the electrode surfaces. Mechanisms of mass transport may include diffusion, convection and electromigration.

In some circumstances, one of the possible contributing factors to the overvoltage is dominant and it is possible to say that the electrode process is, for instance, 'mass transport' or 'charge transfer' limited. In other cases, more than one step is responsible for significantly restricting the current flow, and overvoltage contributions from all such steps must be taken into account. Polarization losses of various kinds can sometimes be distinguished by subjecting the electrode to current or voltage transients (e.g. step or sinusoidal functions) and analysing the response.

Charge transfer kinetics

As noted above, the charge transfer step can involve either electron transfer, as in the case, say, of a $Pt(s)|Fe^{3+}(aq),Fe^{2+}(aq)$ electrode, or ion transfer, as in the case, say, of a $Zn(s)|Zn^{2+}(aq)$ system. While the theoretical treatment of the two forms of charge transfer are essentially different in nature, the final equations relating current and voltage turn out to be very similar.

In the case of redox electrodes, the ease with which electrons can tunnel through a potential barrier of the type present at an electrode interface makes the use of classical activated complex theory (with the electrons as one reactant) inappropriate. In Fig. 2.11(a) an electron energy diagram of a redox electrode at equilibrium is shown. For an electron transfer between the phases to be successful, it is necessary for the acceptor or donor in solution to have an energy level exactly equal to a complementary level in the metal. In the equilibrium situation it is seen that there is an equal chance of transfer of an electron from a filled metal level to an unoccupied

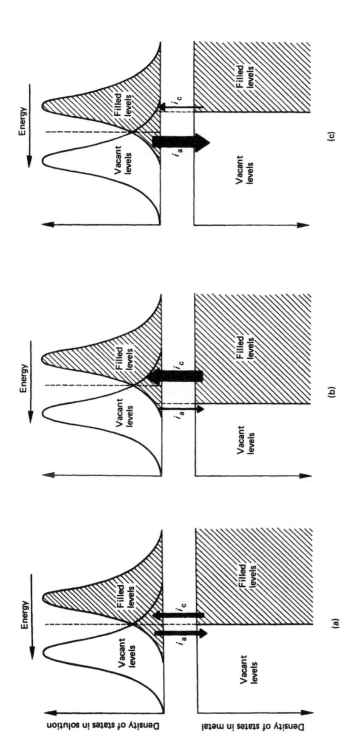

Fig. 2.11 Occupation of energy levels in a metal and a solution redox couple

solution level of the same energy or from an occupied solution level to an empty metal level. By making the electrode potential more negative (Fig. 2.11b), the Fermi level of the metal is raised, more filled electronic states in the metal overlap with unfilled states in the solution and hence there is a relatively greater tendency for electron transfer from the metal to the solution. The opposite situation occurs when the electrode potential is made more positive (Fig. 2.11c). The current flowing at a redox electrode is not limited by the rate at which electrons can tunnel to or from the metal; rather it is restricted by the rates at which the solution species can undergo thermal fluctuations to produce transient species (e.g. ions with abnormal hydration sheaths, or molecules with stretched bonds) which have suitable electronic states. For the electrode process

$$Ox(aq) + e \rightleftharpoons Red(aq)$$

the forward and reverse reactions can be considered as normal heterogeneous rate processes, but with rate constants which are potential dependent. Thus

$$i_c = k_c FA[Ox] \tag{2.48}$$

$$i_a = k_a FA[Red] \tag{2.48'}$$

where k_c and k_a are heterogeneous rate constants. A is the area of the electrode/solution interface and $[Ox]$ and $[Red]$ are the surface concentrations of the electroactive species – here assumed to be the same as the bulk values. The rate constants have the form

$$k_c = k_o \exp \left[-\alpha F(E - E^\ominus)/RT \right] \tag{2.49}$$

and

$$k_a = k_o \exp \left[(1 - \alpha)F(E - E^\ominus)/RT \right] \tag{2.50}$$

where E is the potential of the electrode, $E^\ominus$ is its standard equilibrium potential and k_o is the single heterogeneous rate constant characteristic of the redox system at the temperature of interest. α is a constant between 0 and 1, known as the transfer coefficient or symmetry factor; its value is generally close to 0.5.

For the electrode to be in equilibrium,

$$
\begin{aligned}
i_o &= i_c \ (= i_a) \\
&= k_o FA \, [Ox] \exp \left[-\alpha F(E - E^\ominus)/RT \right] \\
&= k_o FA \, [Ox] \exp \ln \left[\frac{[Ox]^{-\alpha}}{[Red]^{-\alpha}} \right] \\
&= k_o FA \, [Ox]^{(1-\alpha)} [Red]^{\alpha} \tag{2.51}
\end{aligned}
$$

This general expression for the exchange current reduces to

$$i_o = k_o FA[Ox] = k_o FA[Red] \tag{2.52}$$

for the situation where [Ox] = [Red], i.e. at the standard potential of the couple, and to

$$i_o = k_o FA \qquad (2.53)$$

for the special case corresponding to the situation where

$$[Ox] = [Red] = 1$$

Whenever there is a net flow of current, this is taken by convention as

$$i = i_c - i_a$$

For $i_c \neq i_a$, it is no longer strictly correct to equate the surface and bulk concentrations of the electroactive species. However, if the charge transfer rate is current limiting, then as a first approximation it can be assumed that the surface concentration does not deviate significantly from the bulk value. Hence

$$i = k_o FA \ \{[Ox] \exp [-\alpha F(E - E^\ominus)/RT] - [Red] \exp [(1 - \alpha)F(E - E^\ominus)/RT]\}$$

Also $\eta = E - E_{eq}$, so that

$$(E - E^\ominus) = \eta + \frac{RT}{F} \ \ln \ \frac{[Ox]}{[Red]}$$

Hence on substituting for $(E - E^\ominus)$ and recalling the expression derived above for i_o,

$$i = i_o \left[\exp \frac{-\alpha F\eta}{RT} - \exp \frac{(1-\alpha)F\eta}{RT} \right] \qquad (2.54)$$

This equation is known as the Butler–Volmer relationship. Figure 2.12 shows that the net current flowing at an electrode can be considered as the

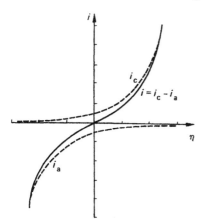

Fig. 2.12 Net current flow, i, at an electrode as the resultant of cathodic and anodic contributions i_c and i_a

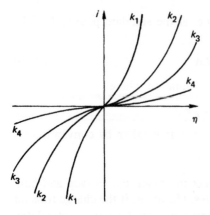

Fig. 2.13 Influence of k_o on the shape of the current–potential curve, for constant α, $k_1 > k_2 > k_3 > k_4$

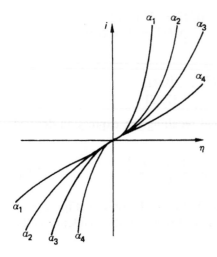

Fig. 2.14 Influence of α on the shape of the current–potential curve, for constant k_o

resultant or sum of cathodic and anodic contributions. Note that in eq. (2.54) and in subsequent equations dealing with electrode kinetics, the current is a direct function of the electrode area, A. It is often useful to normalize such equations with respect to area by replacing the current i by the current density i/A.

The exact shape of the curve is defined by the two parameters k_o and α. In Fig. 2.13, α is held constant at 0.5 while k_o is varied. It can be seen that if the heterogeneous rate constant is high, large currents can be generated by very small overvoltages. In Fig. 2.14 the variation of α for a constant k_o is seen to alter the symmetry of the current/voltage behaviour. An interesting feature of this figure is the coincidence of all the curves near the origin: use can be made of this for determining i_o. The polarization resistance, R_p, of the electrode is defined as the inverse of the current/potential gradient at the origin, i.e.

$$R_p = 1/(di/d\eta)_{\eta=0} \tag{2.55}$$

Now

$$\frac{di}{d\eta} = i_o \left[\left(\frac{-\alpha F}{RT} \exp \frac{-\alpha F \eta}{RT} \right) - \left(\frac{(1-\alpha)F}{RT} \exp \frac{(1-\alpha)F\eta}{RT} \right) \right]$$

Therefore

$$\left(\frac{di}{d\eta} \right)_{\eta=0} = i_o \left[\frac{-\alpha F}{RT} - \frac{(1-\alpha)F}{RT} \right] = \frac{-i_o F}{RT}$$

Hence

$$R_p = - RT/i_o F \tag{2.56}$$

i_o can therefore be found by taking the tangent to the current/potential curve at the origin. In practice, when η is small, the general expression for the current can be simplified by using the relationship:

$$\exp(-x) \approx 1 - x \quad \text{for } x \text{ small}$$

Under these circumstances,

$$i = - i_o F \eta / RT \tag{2.57}$$

Another construction for obtaining i_o is suggested later.

There are a number of methods available for determining α. For instance, rearranging eq. (2.51) gives

$$i_o/[\text{Ox}] = k_o FA([\text{Ox}]/[\text{Red}])^{-\alpha}$$

so that

$$\ln i_o/[\text{Ox}] = \ln k_o FA - \alpha \ln [\text{Ox}]/[\text{Red}]$$

Hence a plot of $\ln i_o/[\text{Ox}]$ versus $\ln [\text{Ox}]/[\text{Red}]$ gives α from the slope. The transfer coefficient may also be evaluated using a form of the Tafel equation as outlined below.

If α and i_o are known, then k_o is given as

$$k_o = (i_o/FA) [\text{Ox}]^{(1-\alpha)} [\text{Red}]^{\alpha} \tag{2.58}$$

in the general case, or

$$k_o = (i_o/FA) [\text{Ox}] = (i_o/FA)[\text{Red}] \tag{2.59}$$

at the standard potential of the couple, or

$$k_o = i_o/FA \tag{2.60}$$

where

$$[\text{Ox}] = [\text{Red}] = 1$$

For large absolute values of η, only that component of the electrode process favoured by the direction of the overpotential need be considered, i.e. the 'back reaction' can be neglected. From the definitions of i_c and i_a we have

$$i_c/i_a = \exp(-F\eta/RT) \tag{2.61}$$

or

$$\ln(i_c/i_a) = -\eta/0.0256 \quad \text{at } 25°C$$

Thus for an overvoltage of -118 mV, $i_c/i_a = 100/1$. For overvoltages of 200 mV and above, only the cathodic or anodic contribution to the total current need be considered. For a large cathodic overvoltage,

$$i = i_c - i_a \approx i_c$$
$$= i_o \exp(-\alpha F\eta/RT)$$

Therefore

$$\ln i = \ln i_o - \alpha F\eta/RT \tag{2.62}$$

Plots of $\ln i$ as a function of η are shown in Fig. 2.15. By extending the linear portions of these curves to $\eta = 0$, a value of the exchange current is given, while the α coefficients may be evaluated from their gradients. Equation (2.62) may be recast in the form:

$$\eta = (RT/\alpha F) \ln i_o - (RT/\alpha F) \ln i$$

or as

$$\eta = A + B \ln i$$

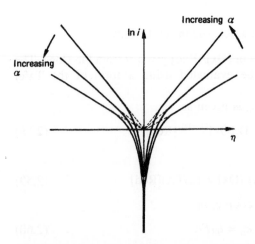

Fig. 2.15 Plots of $\ln i$ as a function of η: extrapolation of the linear portions of the curves to $\eta = 0$ gives the value of the exchange current i_o, while α may be evaluated from their gradients

which is the usual form of the well-known Tafel equation. A similar expression exists for anodic currents.

The electric field which actually affects the charge transfer kinetics is that between the electrode and the plane of closest approach of the solvated electroactive species ('outer Helmholtz plane'), as shown in Fig. 2.2. While the potential drop across this region generally corresponds to the major component of the polarization voltage, a further potential fall occurs in the 'diffuse double layer' which extends from the outer Hemlholtz plane into the bulk of the solution. In addition, when ions are specifically absorbed at the electrode surface (Fig. 2.2c), the potential distribution in the inner part of the double layer is no longer a simple function of the polarization voltage. Under these circumstances, serious deviations from Tafel-like behaviour are common.

Many redox reactions at electrodes involve transfer of more than one electron. It is agreed that such processes usually involve several consecutive one-electron steps rather than a simultaneous multi-electron transfer. The kinetics of the overall reaction (and hence the current flowing) are complicated by such factors as the lifetimes of the transient intermediate species.

The theoretical model generally used for predicting the overvoltage–current function for metal/metal ion systems is based on the quasi-thermodynamic arguments of transition state theory. The anodic charge transfer process is considered to involve the rupture of the bond between an 'adatom' – i.e. a metal atom in a favourable surface site – and the metal, followed by, or coincident with, the formation of electrostatic bonds between the newly formed ion and solvent or other complexing molecules. The cathodic charge transfer follows this mechanism in reverse:

- anodic charge transfer: $M(s) \rightleftharpoons M(ad) \rightarrow M^{n+} (aq) + ne$
- cathodic charge transfer: $M^{n+} (aq) + ne \rightarrow M(ad) \rightleftharpoons M(s)$

It is assumed that for pure charge transfer current limitation, the equilibrium between $M(s)$ and $M(ad)$ is fast and potential independent. Under these conditions the relevant equations are

$$k_a = k_o \exp [(1 - \alpha)nF(E - E^\ominus)/RT] \tag{2.63}$$

$$i_a = k_o nFA[M(ad)]\exp[(1 - \alpha)nF(E - E^\ominus)/RT] \tag{2.64}$$

$$k_c = k_o \exp\{- \alpha nF(E - E^\ominus)/RT\} \tag{2.65}$$

$$i_c = k_o nFA[M^{n+} (aq)]\exp\{- \alpha nF(E - E^\ominus)/RT\} \tag{2.66}$$

$$i_o = k_o nFA[M^{n+} (aq)]^{(1-\alpha)}[M(ad)]^\alpha \tag{2.67}$$

and
$$i = i_o \left[\exp \frac{- \alpha nF\eta}{RT} - \exp \frac{(1-\alpha)nF\eta}{RT} \right] \tag{2.68}$$

The close similarity between eqs (2.68) and (2.62) can be seen. Equation (2.68) leads in the same way to Tafel behaviour at high overvoltages.

Concentration overvoltage

The term 'concentration overvoltage' is used to describe restrictions in the current caused by concentration changes of the electroactive species at the electrode surface. In practical batteries, such an effect is only important at relatively high current densities, although with 'immobilized' electrolyte phases which are present in polymer electrolyte-based cells (Chapter 7), convection is absent and concentration polarization can be a serious problem. On the other hand, in cells using ceramic solid electrolytes where the transport number of the mobile ion is unity, there is no concentration overvoltage even at high rate. If attention is focused first on a liquid electrolytic phase, it is seen that the supply of electroactive molecules or ions to the electrode surface can be maintained either by some process of mass transport from the bulk of the phase or by a (potential independent) chemical reaction occurring in the electrolytic phase near the interface. The principal mechanisms of mass transport in batteries are diffusion in a concentration gradient and migration in an electric field. If ions are present in the electrolyte which are not involved in the electrode process, then transport by electromigration may be small. On the other hand, when the electroactive materials are responsible for carrying all the current in the electrolyte phase, it is not possible to distinguish mass transport polarization and ohmic potential drop. For an electrode process $Ox + ne \rightarrow Red$, concentration overvoltage is given by

$$
\begin{aligned}
\eta &= E - E_{eq} \\
&= \left\{ E^{\ominus} + \frac{RT}{nF} \ln \frac{[Ox]_o}{[Red]_o} \right\} - \left\{ E^{\ominus} + \frac{RT}{nF} \ln \frac{[Ox]}{[Red]} \right\} \\
&= \frac{RT}{nF} \ln \left\{ \frac{[Ox]_o}{[Ox]} \cdot \frac{[Red]}{[Red]_o} \right\}
\end{aligned}
\tag{2.69}
$$

where $[Ox]_o$ and $[Red]_o$ represent the concentrations of electroactive species adjacent to the electrode surface as the result of passing a particular level of current.

In the case of mass transport by pure diffusion, the concentrations of electroactive species at an electrode surface can often be calculated for simple systems by solving Fick's equations with appropriate boundary conditions. A well known example is for the overvoltage at a planar electrode under an imposed constant current and conditions of semi-infinite linear diffusion. The relationships between concentration, distance from the electrode surface, x, and time, t, are determined by solution of Fick's second law, so that expressions can be written for $[Ox]_o$ and $[Red]_o$ as functions of time. Thus, for

$$
M^{n+}(aq) + ne \rightarrow M(s)
$$

and assuming that the activity of the deposited metal remains at unity, and that the surface layer is not completely depleted of M^{n+} (aq) ions, then

$$[M^{n+}]_o = [M^{n+}] - \frac{2it^{1/2}}{nFAD^{1/2}\,\pi^{1/2}} \qquad (2.70)$$

where D is the diffusion coefficient of M^{n+} (aq) and i is the constant impressed current. Further

$$E = E^{\ominus} + \frac{RT}{nF}\,\ln\,[M^{n+}]_o \quad \text{and} \quad E_{eq} = E^{\ominus} + \frac{RT}{nF}\,\ln\,[M^{n+}]$$

so that

$$\eta = \frac{RT}{nF}\,\ln\left\{\frac{nFAD^{1/2}\,\pi^{1/2}\,[M^{n+}] - 2it^{1/2}}{nFAD^{1/2}\,\pi^{1/2}\,[M^{n+}]}\right\} \qquad (2.71)$$

However, in most practical situations it is not possible to obtain analytical expressions of this type. A useful model, originally intended for systems with a stirred electrolytic phase, assumes that a thin 'stationary' or 'stagnant' layer extends some way from the electrode surface into the bulk electrolytic phase and that beyond this layer, no significant concentration gradients exist (Fig. 2.16). This model predicts that after polarization for an initial period, a time-invariant overvoltage will be established. This model has been applied with some success to unstirred systems where it is thought that natural convection caused by density variations (arising from changes in the composition of the electrolyte due to the electrode reaction) permits a similar layer system to be established. Such a model is not applicable where the electrolyte is immobilized, nor is it useful for electrodes with rough or porous surfaces.

Again, for the process M^{n+} (aq) + $ne \rightarrow$ M(s) and assuming first that diffusion is the only mechanism of transport across the stationary layer, in the steady state, the flux of electroactive species across the layer must be balanced by the flux of charge at the electrode surface. For a linear concentration gradient across a stationary layer of thickness δ, application of Fick's first law gives

$$i = nFAD\,\{([M^{n+}] - [M^{n+}]_o)/\delta\} \qquad (2.72)$$

The form of this equation shows that the concentration gradient can have any value from zero up to a maximum of $[M^{n+}]/\delta$ where the surface

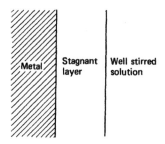

Metal | Stagnant layer | Well stirred solution

Fig. 2.16 Stationary layer model

concentration of M^{n+} (aq) has fallen to zero. This maximum concentration gradient is associated with a maximum, or limiting, current given by

$$i_{lim} = nFAD[M^{n+}]/\delta \tag{2.73}$$

Note that it is straightforward to correct equations such as eqs (2.72) and (2.73) for electromigration. Thus

$$i = nFAD \left\{ \frac{[M^{n+}] - [M^{n+}]_o}{\delta(1 - t_{M^{n+}})} \right\} \tag{2.74}$$

and

$$i_{lim} = nFAD \left\{ \frac{[M^{n+}]}{\delta(1 - t_{M^{n+}})} \right\} \tag{2.75}$$

where $t_{M^{n+}}$ is the transport number of M^{n+}(aq) in solution.

The diffusion overvoltage according to this model can be evaluated using the relationship

$$[M^{n+}]_o = (1 - i/i_{lim})[M^{n+}] \tag{2.76}$$

so that

$$\eta = \frac{RT}{nF} \ln \left(\frac{i_{lim} - i}{i_{lim}} \right) \tag{2.77}$$

It should be remembered that in the case of a metal/metal ion electrode, the current can only be mass-transport limited in the cathodic direction. For net anodic currents, an accumulation of metal ions occurs at the electrode surface. The anodic current can greatly exceed the limiting cathodic current (Fig. 2.17), and under these circumstances equations can be simplified to give

$$\eta = \frac{RT}{nF} \ln \frac{|-i|}{i_{lim}} \tag{2.78}$$

where i refers to the anodic current and i_{lim} to the limiting cathodic current.

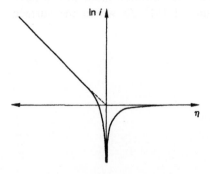

Fig. 2.17 Current–voltage curve for a metal–metal ion electrode

Hence

$$\eta = \left[\frac{RT}{nF} \ln i_{\text{lim}} \right] - \frac{RT}{nF} \ln i \qquad (2.79)$$

This equation has exactly the same form as the Tafel equation: a linear Tafel plot does not necessarily imply a charge-transfer limited process.

In the models for mass transport considered so far, it has been assumed that the electrode surface was smooth. In practice, rough or porous structures are commonly used in battery systems to increase the 'real' surface area and so reduce the charge transfer overvoltage. When surface irregularities are of the same order of magnitude as the thickness of the diffusion layer δ, calculation of diffusion flux becomes very difficult. With porous electrodes the real surface area may be 10^3–10^5 times greater than the projected or geometric surface. However, the presence of a pore structure imposes fundamental changes both on charge-transfer kinetics because of the altered potential distribution and on mass-transport control because of the effect of the pores on concentration gradients. A number of simple models have been studied (e.g. a sequence of parallel cylindrical pores), and while there are no general solutions, a number of limiting cases have been described quantitatively.

In addition to mass transport from the bulk of the electrolyte phase, electroactive material may also be supplied at the electrode surface by homogeneous or heterogeneous chemical reaction. For example, hydrogen ions required in an electrode process may be generated by the dissociation of a weak acid. As this is an uncommon mechanism so far as practical batteries are concerned (but not so for fuel cells), the theory of reaction overvoltage will not be further developed here. However, it may be noted that Tafel-like behaviour and the formation of limiting currents are possible in reaction controlled electrode processes.

The discussion of concentration polarization so far has centred on the depletion of electroactive material on the electrolyte side of the interface. If the metal deposition and dissolution processes involve metastable active surface atoms, then the rate of formation or disappearance of these may be the critical factor in the overall electrode kinetics. Equation (2.69) can be rewritten for 'crystallization overvoltage' as

$$\eta = \frac{RT}{nF} \ln \frac{[\text{M(ad)}]_{\text{eq}}}{[\text{M(ad)}]} \qquad (2.80)$$

where $[\text{M(ad)}]_{\text{eq}}$ and $[\text{M(ad)}]$ represent the surface concentrations (or, better, activities) of M adatoms on the metal surface under conditions of equilibrium and current flow respectively. Some features associated with electrocrystallization and electrodissolution are shown in Fig. 2.18. Atoms such as A are the adatoms; they may be considered as the species formed or lost during charge transfer. (In other theories, adatoms are considered to have undergone only partial charge transfer.) Adatoms are free to move about on surfaces such as XY by an activated process of surface diffusion.

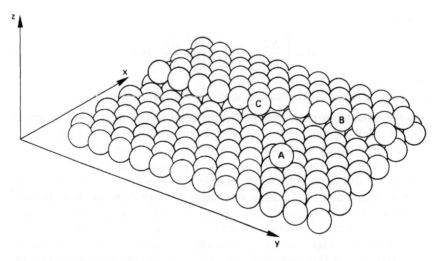

Fig. 2.18 Electrocrystallization on a metal surface. The upper layer of atoms grows by surface diffusion of 'adatoms' (A) across plane *x–y* until they encounter a 'step site' of lower energy, such as B. The adatom may now move along the step until it is located at the even lower energy 'kink site' at C

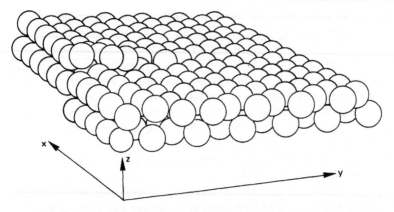

Fig. 2.19 Electrocrystallization on a metal surface. Growth can proceed continuously in the *z*-direction as the step rotates around the screw dislocation

Atoms at B and C are seen to exist at different positions of lower energy known as step sites.

At low current densities, electrocrystallization is thought to proceed by incorporation of adatoms in step sites. The steps may then grow, and provided that each step is located at a screw dislocation (Fig. 2.19) the number of step sites remains constant during deposition or dissolution. At higher current densities, surface nucleation occurs, thus greatly increasing the number of growth sites (Fig. 2.20). At high anodic current densities, a different form of surface nucleation must be considered: the formation of cavities, one atom deep. When discussing electrocrystallization or electro-

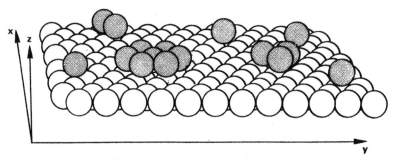

Fig. 2.20 Electrocrystallization on a metal surface. At high concentrations of adatoms two-dimensional surface nuclei may form which provide growth steps

dissolution overvoltage, it is assumed that the phase transfer process itself does not restrict the current. The limiting process is the supply or removal of M adatoms on the surface. This in turn may be affected by the rate of surface diffusion, the rate of interaction of adatoms with growth steps, the concentration and distribution of nuclei, etc. The overvoltage may therefore follow a wide variety of current functions. If surface diffusion is rate-limiting, then in the simplest cases the overvoltage/current relationship becomes indistinguishable from that of charge transfer. This is also the case for sufficiently large rates of surface nucleus formation. Where both nucleation and screw dislocation growth proceed simultaneously, the current/voltage relationship is complex.

It is therefore seen that eqs (2.42) and (2.43) for the on-load cell voltages during discharge and charge, respectively, can be expanded to give

$$E_{cell} = E_{thermodynamic} - E_{pol}$$
$$= E_{thermodynamic} - \eta_{cathode} - \eta_{anode} - iR$$

and

$$E_{cell} = E_{thermodynamic} + E'_{pol}$$
$$= E_{thermodynamic} + \eta'_{cathode} + \eta'_{anode} + iR$$

The cathodic and anodic overvoltages depend on the dominant rate-limiting processes for a particular current density and direction. For most practical cells, however (and as noted below in Fig. 2.22), low current charge or discharge may be carried out close to the OCV. At higher currents, but within the normal rated values of the cell, the internal iR drop is generally the most significant form of polarization. Only at the highest rates does concentration polarization become dominant.

2.5 Battery characteristics and performance criteria

A cell may be characterised in terms of (1) its available capacity, (2) its available energy, and (3) the power it can deliver. It is not very useful to consider efficiency in the sense used to assess heat engines since batteries cannot be regarded as thermal converters. Rather, the efficiency of an electrochemical power source can be expressed in terms of capacity or of energy delivered.

Capacity

The theoretical capacity of a cell or half-cell may be calculated as

$$Q_T = x(nF) \tag{2.81}$$

where x is the theoretical number of moles of reaction associated with the complete discharge of the cell, using the *lower* of the anode or cathode quantities to calculate xnF if the cell is not balanced. The practical capacity or actual number of coulombs or ampere hours delivered, Q_P, is lower than Q_T if utilization of electroactive material is not 100% due, for instance, to some chemical reaction occurring in the cell which consumes some of the reactants. The rated capacity is the practical capacity of a cell which has been discharged under prescribed conditions until the cell voltage has fallen to a pre-selected cut-off voltage.

The coulombic efficiency of a cell is defined as Q_P/Q_T. It is often more useful to determine the capacity of each half-cell separately, since for operational reasons, most practical batteries do not have an equal number of equivalents of anodic and cathodic reactants.

For purposes of comparison, it is convenient to calculate the specific capacity defined as the capacity divided by the mass of the cell or half-cell, and usually given in units of Ah/kg. In some cases a volume-based specific capacity is preferred (e.g. Ah/dm^3).

Energy

The theoretical available energy for 1 mole of reaction (not for complete discharge) is given by

$$\mathscr{E}_T = - \Delta G = nFE_{cell} \tag{2.82}$$

where E_{cell} is the emf. The actual amount of energy delivered for 1 mole of reaction, or practical available energy, is

$$\mathscr{E}_P = \int_0^{nF} E.dq = \int_0^t (Ei)dt \tag{2.83}$$

and is dependent on the manner in which the cell is discharged. The units of energy are either joules (i.e. watt seconds) or more commonly watt hours (1 Wh = 3600 J). The total energy of a cell is sometimes rather confusingly termed the 'watt hour capacity'. As discussed above, the cell voltage E deviates progressively from its (maximum) thermodynamic value as the rate of discharge increases. Hence the energetic efficiency $\mathscr{E}_P/\mathscr{E}_T$ is a variable quantity, which must be associated with closely defined discharge conditions if it is to be meaningful.

Theoretical and practical energies can also be (and are more usually) expressed in terms of the complete discharge of a particular cell:

$$\mathcal{E}'_T = x(nFE_{cell}) \tag{2.82'}$$

and

$$\mathcal{E}'_P = \int_0^{xnF} E.dq \tag{2.83'}$$

where x is again the number of moles of reaction associated with the completed discharge.

The energy density (also known as specific energy) is the parameter used when assessing relative cell performance. Thus a small battery, weighing 25 g and capable of delivering 40 kJ or 0.012 kWh at a particular discharge current, would be said to have an energy density of 480 Wh/kg. Again it is sometimes more useful to consider a volume based specific energy (i.e. Wh/dm^3).

Power

The level of discharge current drawn from a cell is determined principally by the external load resistance. The power delivered, P, is given by the product of the current flowing and the associated cell voltage:

$$P = iE \tag{2.84}$$

The power rating of a battery specifies whether or not it is capable of sustaining a large current drain without undue polarization. As more and more current is drawn from a cell, the power initially rises; it reaches a maximum and then drops as the cell voltage falls away due to polarization effects (Fig. 2.21). The maximum power point is best determined experi-

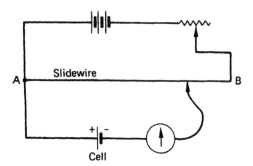

Fig. 2.21 The power of a typical electrochemical cell as a function of discharge current

mentally by measuring E as a function of i. In certain circumstances it is possible to calculate the maximum power point, e.g. if electrode polarization is small and the internal resistance of the cell is known. The rated power of a battery is the power delivered (in watts) under stated discharge conditions.

The rated power density (or specific rated power) is again the most convenient parameter for comparing different battery systems, and is generally quoted in W/kg or W/dm^3. The rated power density and rated energy density are both critical factors when assessing batteries for applications such as motor traction where the battery itself has to be transported.

A cell generally has a maximum permitted continuous power level: prolonged discharge above this value is liable to cause overheating and a consequent degradation of the cell, as discussed below. On the other hand, cells may be capable of a much higher rated instantaneous power. For short discharges the battery does not reach a thermal stationary state and its thermal capacity may be able to accommodate the heat generated if the thermal conductivities of its constituents are sufficiently high. Further, mass transport polarization effects are of less significance for short discharges. Certain batteries, such as those used for starting large diesel engines, depend specifically on their high instantaneous power capabilities.

As will be seen later, cells employing the same chemical system can be designed either for high power or high capacity.

Polarization and discharge curves

More detailed information on the characteristic behaviour of cells under load and at various stages of discharge can be conveyed by graphs. Plots of cell voltage against current are usually referred to as polarization (or performance) curves; graphs of cell voltage as a function of the fraction of discharge completed are known as discharge curves.

A typical cell polarization curve (Fig. 2.22) has three regions. Initially

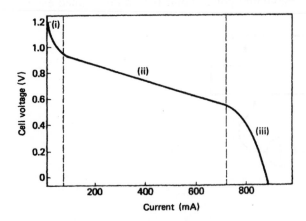

Fig. 2.22 Typical polarization curve for an electrochemical cell

there is a rapid fall in cell voltage at low current drain, due to electrode polarization overvoltage, usually associated to a large extent with one of the two electrode processes. (In a few cases where a passive film on one of the electrodes initially interferes with the discharge, the initial stages may be associated with a rise in cell voltage as the film is destroyed – see Chapter 5.)

In the second, almost linear, region, the internal resistance of the cell components causes a further voltage loss (the 'iR polarization'). In the final region, at relatively high current drain, the iR polarization is combined with further electrode polarization caused by depletion of electroactive materials at the electrode surfaces. While such curves provide much useful information, they must be interpreted with some care. First, it is often necessary to examine polarization curves at various stages of discharge since both electrode and electrolyte characteristics can alter considerably as discharge products build up. Second, polarization curves are usually time-dependent, so that the details of how they have been determined must be examined; e.g. has there been a relatively fast sweep of voltage or current through the requisite values, or has a slow, pseudo-steady-state approach been taken?

Discharge curves show either the open circuit voltage of a cell or half-cell as a function of the fraction of discharge completed, or, much more commonly, the cell voltage during a deep discharge, usually under a fixed load or at constant current. The abscissa may be calibrated in terms of the quantity of electricity passed (Ah) or as a percentage of the theoretical capacity.

A useful method of characterizing the discharge (or charge) rates used either in recording discharge curves or for determining practical capacities is to standardize the current in terms of the nominal cell capacity. A C-rate of τ implies that the nominal capacity of the cell (measured in Ah) is delivered in $1/\tau$ hours; e.g. for a 2 Ah cell, discharge at C/5 signifies a current drain of 0.4 A.

Discharge curves for typical cathodic half-cells are shown in Fig. 2.23. In curve (a) reactants and products are both in liquid or solid solution, while curve (b) represents the case where they form two distinct solid phases. In Fig. 2.24, a two-stage discharge takes place, where all three electroactive components form separate solid phases and the electrolytic phase remains virtually invariant, e.g.

$$2 \; CuS(s) + 2e \rightarrow Cu_2S(s) + S^{2-} \; (s) \qquad (2.85)$$

and
$$Cu_2S(s) + 2e \rightarrow 2Cu(s) + S^{2-} \; (s) \qquad (2.86)$$

A number of empirical equations have been suggested (e.g. Shepherd formulae) for certain types of cell which give cell voltage as a function of discharge depth, current, etc., once a number of constants characteristic of the cell discharge have been evaluated.

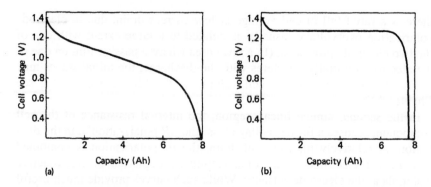

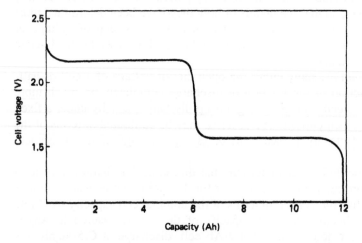

Fig. 2.23 Discharge curves for typical cathodic half-cells. (a) Reactants and products in solution phase, e.g. $Fe^{3+}(aq)$ and $Fe^{2+}(aq)$; (b) reactants and products both form solid phases, e.g. $Ag_2O(s)$ and $Ag(s)$

Fig. 2.24 Two stage discharge curve where all three electroactive components form separate solid phases, as in the cell $Li(s)|LiClO_4$, $PC|CuS(s)$

Battery testing and specifications

The useful life of a practical primary battery is determined principally by the nature of its discharge pattern. Thus the best way of assessing a system for some particular application is to subject it to a discharge which simulates the service conditions. Tests have therefore been developed which recognize the principal function of various types of battery and specify the generation of intermittent or continuous currents of appropriate levels. Such procedures have been standardized for batteries of uniform size and cell configuration by bodies such as the International Electrotechnical Commission (IEC) and the American National Standards Institute (ANSI). New test routines are continually being devised to keep pace with new commercial battery developments and applications. Other tests and specifications have been detailed by various military, technical and commercial organizations.

Most battery tests are based simply on cell discharge through fixed resistors (with or without interspersed recovery periods). From the times required to reach predetermined cut-off voltages, estimates of the practical capacity, Q_P, and of the practical energy, $\mathscr{E}'_P$ can be given.

Other routine tests made on commercial systems include storage (shelf life) tests, assessment under conditions of environmental and mechanical stress, etc. The behaviour of cells under conditions of continuous short circuit and after complete discharge must also be characterized. Testing procedures for particular cell systems are outlined in subsequent chapters.

Rechargeable systems

Spontaneous cell reactions can often be reversed by applying a potential difference to the cell in such a way as to drive current through it in the opposite direction to the discharge. In some cases the reverse current will regenerate the original cell components only to a limited extent; the remainder of the current is consumed in unwanted side reactions. While it is not in general possible to distinguish 'primary' and 'secondary' systems in any absolute way, a number of completely irreversible electrode systems do exist, for example metals with low hydrogen overvoltage such as magnesium in aqueous electrolytes. It is impossible to recharge a $Mg(s)|Mg^{2+}(aq)$ electrode since hydrogen evolution is both thermodynamically and kinetically preferred to the reduction of magnesium ions. The voltage of a cell under charge is higher than its emf (eq. 2.43). In some systems when recharging at constant current is almost complete, the voltage rises sharply as new overcharge reactions commence at the electrodes; in others there is no obvious break. For practical batteries, appropriate charging regimes have been derived for each particular system: details of some of these are given in Appendix 1.

The ability of a cell to accept charge or be (re-)charged is measured in terms of the capacity and energy efficiencies of the charge/discharge cycle. The capacity or ampere-hour efficiency of a cell cycled under stated conditions of rate and depth is defined as

$$\frac{\int_0^t i_{dis}\, dt}{\int_0^t i_{ch}\, dt}$$

where i_{dis} and i_{ch} refer to the currents flowing during discharge and charge, respectively. The inverse of the capacity efficiency is known as the charge factor. The charge factor is substantially greater than unity if significant side reactions occur while the cell is being charged. The overall cycle energy efficiency is given by

$$\frac{\int\limits_0^t E_{\text{dis}}\, i_{\text{dis}}\, \mathrm{d}t}{\int\limits_0^t E_{\text{ch}}\, i_{\text{ch}}\, \mathrm{d}t}$$

where E_{dis} and E_{ch} are the cell voltages during discharge and charge, respectively. It is emphasized that the ability of a cell to accept charge is dependent on the way in which the charging operation is carried out and that these efficiency ratings are not absolute quantities. In practice, cycle efficiencies are usually quoted in terms of particular C-rates for charge and discharge to a fixed depth or voltage.

Cycle life is the number of times a cell can undergo a charge/discharge sequence before its performance (as measured by its capacity or energy storage efficiency) has been degraded to below some arbitrary limit. For shallow discharge (say <20%) a much longer cycle life is obtained. Standard tests generally require a discharge to 75% or 80% of rated capacity. Charge retention is a measure of whether the percentage of nominal capacity lost as a function of time due to self-discharge reactions occurring within the cell is large or small. Charge retention is much poorer at high temperatures.

In addition to having a high cycle efficiency, a practical rechargeable cell must meet a number of other criteria. In particular it must have a satisfactory rate of charge acceptance. This is governed by the same factors that control the magnitude of normal discharge currents: exchange current, area and nature of electrode surface, ohmic resistance, etc. The latter is of special importance in charging processes, since if high charging voltages are applied the problem of overheating and its effect on the chemical and mechanical stability of the system may become critical. Second, if solid phases are formed during recharge, their morphology must be suitable: dimensionally stable, coherent, non-dendritic (especially with metallic phases, lest internal short circuits occur), etc.

Because of competing side reactions, the full charge capacity of many secondary battery systems can only be achieved by overcharging, i.e. by passing charge in excess of the theoretical capacity during the recharge. In aqueous cells the side reactions associated with the excess charge usually involve electrolysis to form hydrogen or oxygen. The nature of side reaction products can sometimes be controlled by regulating the relative capacity of anode and cathode. If a gas is formed, provision must be made for its venting or, as in the case of sealed cells, recombination.

Certain problems sometimes arise with rechargeable multicell batteries which are associated with non-uniform capacities of the individual cells. In charging such batteries there must be either overcharge in some cells or undercharge in the whole system. More importantly, on deep discharge, cell reversal may take place where the lowest capacity cell suffers a polarity reversal. With some, but not all, systems such an occurrence is catastrophic and a number of electrochemical and electrical reversal protection devices have been incorporated in practical rechargeable multicell assemblies.

Thermal management

'Thermal management' is defined as the method of operation, for a particular engineering design, whereby a battery system is maintained within a specified temperature range while undergoing charge or discharge processes. Cells are generally characterized by a maximum working temperature, above which corrosion and other irreversible destructive processes are unacceptably rapid; cells also have, in practice, a minimum working temperature below which the electrolyte has too high a resistance or is liable to undergo a phase change – this is of most importance in 'high temperature' cells (Chapter 8).

The thermal energy generated or absorbed by an electrochemical cell is determined first by the thermodynamic parameters of the cell reaction, and second by the overvoltages and efficiencies of the electrode processes and by the internal resistance of the cell system. While the former are generally relatively simple functions of the state of charge and temperature, the latter are dependent on many variables, including the cell history.

Overheating problems are most common in large batteries (i.e. those with low surface to volume ratios) undergoing high rates of charge or discharge. Some of the thermal energy generated by the cell process may be dissipated as heat to the surroundings* either by natural air convection, possibly augmented by cooling fins, channels or gaps between the cells, etc., or by use of heat exchangers involving forced circulation of gas or liquid coolant or pumping of the electrolyte through an external cooling unit. The remaining energy remains within the cell, resulting in an increase in temperature. The continuous rated power level of a battery is set at such a value that the steady state temperature does not exceed the maximum working temperature.

The thermal management problems of high temperature batteries are more complex. First, a heating system is required to raise the temperature of the battery to the operating value, or to maintain it at this value when no current is being drawn; second, an insulating container must be provided to minimize heat loss and thus raise the overall energy efficiency; and finally, a cooling system may be necessary to prevent overheating during high rate operation.

With the growing interest in large high rate batteries for traction and load levelling applications, the development of models for heat transfer

---

* Problems concerning heat dissipation in spacecraft batteries arise as a consequence of the absence of convective heat transfer in a vacuum.

analysis has become of increasing importance to design engineers in recent years. Once the thermal properties (heat capacity, thermal conductivities, etc.) are known a number of numerical and analytical models are available for calculating temperature distribution during various operating schedules.

3 Primary aqueous electrolyte cells

Marrio Lazzari✠ and Colin A Vincent

3.1 Introduction

The aqueous electrolyte battery group includes some of the oldest but still most commercially important primary and secondary systems. Thus, despite considerable advances in the design and manufacture of new types of cell over recent years, the classical Leclanché cell and its variants, based on zinc and manganese dioxide, now more than a century old, are still by far the primary cells in largest production. The strong market position of these cells is due to a combination of factors including low cost of materials and ease of manufacture, together with performance characteristics which are suitable for a wide range of practical applications.

With the continuing development of portable electric and electronic equipment, demand for small primary aqueous electrolyte cells continues to expand and a growth rate in world production in the range of 6–12% per annum has been maintained over the past 25 years. It is estimated that there is now an annual consumption averaging 8–15 cells per person in the Western developed countries.

Almost all modern practical aqueous primaries are referred to as 'dry cells'. This designation should not be confused with the rather specialized 'solid state cells' which make use of the recently discovered true solid electrolytes. Rather, the term implies that the aqueous electrolyte phase has been immobilized by the use of gelling agents or by incorporation into microporous separators. Such procedures permit the cells to operate in any orientation and reduce the effects of leakage should the container become punctured.

The range of commercially available cells is so large that it would be impossible to include here a comprehensive description of their design and performance characteristics. Instead, examples of the more important cells are discussed in detail, together with features of some other systems of

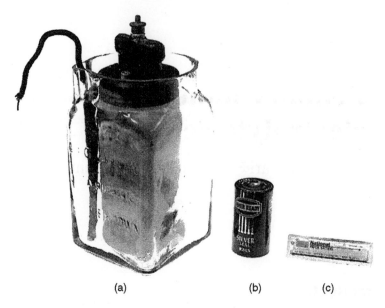

(a)   (b)   (c)

Fig. 3.1 Leclanché cells, past and present: (a) Leclanché wet cell; (b) D-size dry cell (Ever Ready); (c) 'paper battery'. (Matsushita Battery Industrial Co.)

interest which have often been developed with particular applications in mind.

3.2   Leclanché cells

The name 'Leclanché cell' is given to the familiar primary system consisting of a zinc anode, manganese dioxide cathode and an electrolyte of ammonium chloride and zinc chloride dissolved in water. The alternative designation 'zinc–carbon cell' is broader and includes the so-called zinc chloride system which, due to a different electrolyte composition, is characterized by a different discharge mechanism. The Leclanché cell may be written as

$$Zn(s)|ZnCl_2(aq),NH_4Cl(aq)|MnO_2(s),C(s)$$

and the OCV* is in the range 1.55–1.74 V. Despite the many advances which have brought this cell to a high degree of reliability and greatly improved performance, the electrochemical system of the modern cell is basically the same as that proposed by George Leclanché in 1866 (Fig. 3.1a).

The original Leclanché cell was a 'wet cell' assembled in a glass jar. The cathode was a carbon rod or plate immersed in a mixture of manganese

---

* OCV: open circuit voltage – see Glossary

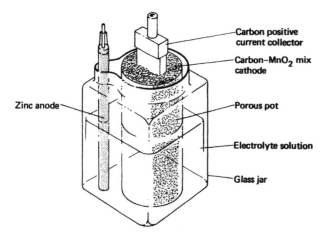

Fig. 3.2 Schematic view of the original Leclanché wet cell

dioxide and carbon powder contained in a porous ceramic pot. The anode was a zinc rod which was immersed in the electrolyte, a saturated solution of ammonium chloride (Fig. 3.2). By 1868 more than 20 000 of these cells were in service in European telegraph systems. Twenty years after the original cell had been proposed, Carl Geissner patented the idea of manufacturing the zinc anode in the form of a cup which would act as a container for the electrolyte (which was immobilized with plaster of Paris). Combination of this development of a zinc container with the use of a manganese dioxide/carbon anode and an electrolyte immobilized using a cereal paste laid the foundations of a modern dry cell technology and led quickly to a massive production of Leclanché cells which had exceeded an annual production of two million units by the turn of the century.

Discharge mechanisms

The discharge mechanism of the Leclanché cell is complex and not all the details are yet fully understood. The basic process consists of oxidation of zinc at the anode to form zinc ions in solution:

$$Zn(s) \rightarrow Zn^{2+}(aq) + 2e \qquad (3.1)$$

together with a reduction of Mn (IV) to a trivalent state as $MnO.OH(s)$ or $Mn_2O_3.H_2O(s)$, at the cathode, e.g.

$$2MnO_2(s) + 2H_2O + 2e \rightarrow 2MnO.OH(s) + 2OH^-(aq) \qquad (3.2)$$

The initial products of the electrode reactions may then undergo further reactions in solution. The prevailing process is

$$Zn^{2+}(aq) + 2OH^-(aq) + 2NH_4^+(aq) \rightarrow Zn(NH_3)_2^{2+}(aq) + 2H_2O \quad (3.3)$$

followed by the formation of the slightly soluble $Zn(NH_3)_2Cl_2$:

$$Zn(NH_3)_2^{2+}(aq) + 2Cl^-(aq) \rightarrow Zn(NH_3)_2Cl_2(s) \qquad (3.4)$$

For light discharges and with certain oxides an alternative reaction is

$$Zn^{2+}(aq) + 2MnO.OH(s) + 2OH^-(aq) \rightarrow ZnO.Mn_2O_3(s) + 2H_2O \quad (3.5)$$

At lower NH_4^+ concentrations, the zinc ions precipitate out as one or more oxychloride species, e.g.

$$5Zn^{2+}(aq) + 2Cl^-(aq) + 8OH^-(aq) \rightarrow ZnCl_2.4Zn(OH)_2(s) \qquad (3.6)$$

or as the hydroxide:

$$Zn^{2+}(aq) + 2OH^-(aq) \rightarrow Zn(OH)_2(s) \qquad (3.7)$$

The principal overall cell reactions can therefore be summarized as

$$Zn(s) + 2MnO_2(s) + 2NH_4Cl(aq) \rightarrow$$
$$2MnO.OH(s) + Zn(NH_3)_2Cl_2(s) \qquad (3.8)$$

and

$$Zn(s) + 2MnO_2(s) \rightarrow ZnO.Mn_2O_3(s) \qquad (3.9)$$

However, if the initial ammonium chloride concentration is low, then the following processes may be better approximations to the actual cell reaction:

$$4Zn(s) + ZnCl_2(aq) + 8MnO_2(s) + 8H_2O(1) \rightarrow$$
$$8MnO.OH(s) + ZnCl_2.4Zn(OH)_2(s) \qquad (3.10)$$

and

$$Zn(s) + 2MnO_2(s) + 2H_2O(1) \rightarrow 2MnO.OH(s) + Zn(OH)_2(s) \quad (3.11)$$

Calculations of the emf of cells based on these reactions provide values within the wide range of 1.5–1.7 V, characteristic of undischarged cells formed with different samples of manganese dioxide and electrolyte pH. However, during the discharge of practical cells, inhomogeneities in the solution and cathode phases may produce a much more complicated reaction sequence, as will be discussed below.

Electrolyte

The electrolyte in the original Leclanché cell was a saturated solution of ammonium chloride. The addition of zinc chloride was soon found to be beneficial to cell performance, and since Geissner's time the electrolyte has contained both salts. The ternary phase diagram for the system $ZnCl_2$–NH_4Cl–H_2O at room temperature is shown schematically in Fig. 3.3. The important variables to be considered are the composition, pH and conductivity of the electrolyte. A typical composition for an undischarged cell is: NH_4Cl, 28%; $ZnCl_2$, 16%; H_2O, 56%. However, there are variations depending on manufacturer, and zinc chloride concentrations down to 9%

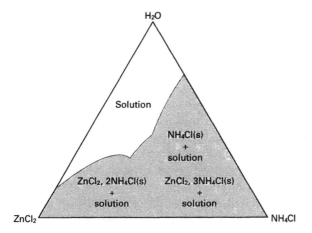

Fig. 3.3 Schematic ternary phase diagram for the system $ZnCl_2$–NH_4Cl–H_2O at room temperature

by weight are common. The solution is often saturated with ammonium chloride to enable the excess to compensate for consumption of this salt during discharge. The zinc chloride content is limited by the formation of the solid phases $ZnCl_2.3NH_4Cl$ and $ZnCl_2.2NH_4Cl$ (as seen in Fig. 3.3) which increase the internal resistance of the cell. The pH of the electrolyte is controlled by the hydrolysis reactions of both the ammonium and the zinc ions, and increases with increasing zinc or ammonium ion concentration.

Since the basic cathode reaction (3.2) involves the production of OH^- ions, the resulting increase in pH causes the formation of ammonia, which in turn complexes with zinc ions. Initially a precipitate of slightly soluble diamminozinc chloride, $Zn(NH_3)_2Cl_2$, is formed, but according to the discharge conditions, as further ammonia is produced this may be converted to the more soluble tetramminozinc chloride, $Zn(NH_3)_4Cl_2$:

$$Zn^{2+}(aq) + 2Cl^-(aq) \xrightarrow{2NH_3} \underset{\text{Slightly soluble}}{Zn(NH_3)_2Cl_2} \xrightarrow{2NH_3}$$
$$Zn(NH_3)_4^{2+}(aq) + 2Cl^-(aq)$$
$$(3.12)$$

The formation of these complexes serves to buffer the rise in pH due to the cathodic discharge process. Precipitation of $Zn(NH_3)_2Cl_2$ has the effect of increasing cell resistance.

In the anodic region, the concentration of Zn^{2+} ions increases as discharge proceeds, leading to a decrease in pH due to hydrolytic reactions of the type

$$Zn^{2+}(aq) + H_2O \rightarrow Zn(OH)^+(aq) + H^+(aq)$$

A pH gradient is thus established during cell discharge.

Cathode

The cathode mix or 'bobbin' consists basically of manganese dioxide to which carbon, in finely divided form (e.g. as acetylene black), has been added in order to provide adequate electronic conductivity and to retain the electrolyte solution in contact with the MnO_2 particles. Graphite, once the normal choice for Leclanché cells, is now used only occasionally for certain high current cells, and has been replaced by acetylene black. The current collector is generally also made of carbon, in the form of either a rod or a thin sheet. The carbon rod in cylindrical cells is treated with wax to reduce its porosity and prevent ingress of oxygen, water loss and electrolyte leakage. In Leclanché cells, in contrast to other zinc–carbon cells, a measure of porosity is retained to permit venting of hydrogen or carbon dioxide and hence to obviate the need for installing a resealable vent with the attendant cost implications. The active material is the MnO_2 and it was realized from early in the development of the Leclanché cell that use of naturally occurring MnO_2 from different sources produced striking variations in cell performance. Even now, when synthetic MnO_2 is used in almost all high performance batteries, slight changes in the method of preparation can have significant effects. MnO_2 has numerous allotropic forms, with subtle crystal structure modifications and a wide variety of surface types. So far it has not proved possible to relate in any exact way the crystal structure, surface properties, etc. of a particular sample of MnO_2 with its corresponding electrochemical behaviour. The only decisive test of the effectiveness of MnO_2 as a battery material is its behaviour in the cell. Much has been written about the 'art' rather than the 'science' of Leclanché cell manufacture!

Naturally occurring MnO_2 ores from a number of sources (and in particular Ghana, Gabon and Mexico), known as NMD, are practically free from significant amounts of heavy metal impurities and can be used directly after washing processes without chemical purification. Improvements in cell performance may be obtained by 'activating' the ore by heating to convert it to Mn_2O_3 and then treating the product with dilute sulphuric acid to reform MnO_2 together with $MnSO_4$. The activated MnO_2, known as AMD, is characterized by a higher surface area.

'Synthetic' MnO_2 is obtained from natural MnO_2 which has been reduced to MnO and thence put into solution using sulphuric acid. The resulting manganous sulphate is purified and then oxidized either chemically to yield CMD or electrochemically to yield EMD, which are MnO_2 phases with high surface area and consistent properties, and which are used in high quality Leclanché cells. CMD and EMD have different crystal structures from the phases existing in the natural ores. They are also free from manganite, $MnOOH$, which in NMD-based cells results in a slightly lower OCV. While more expensive to produce, they provide a higher capacity and better rate capability in Leclanché and other aqueous primary cells.

The overall cathodic reaction (eq. 3.2) in which $MnO_2(s)$ is reduced to $MnO.OH(s)$ is now known to involve a solid state diffusion process in which protons are transported from the surface to the interior of the MnO_2 grains:

$$H_2O(1) \rightleftharpoons H^+(surface) + OH^-(aq)$$

$$H^+(surface) \rightarrow H^+(bulk)$$

$$xMnO_2(s) + H^+(bulk) + e \rightarrow (MnO_2)_{x-1}(MnO.OH)(s) \qquad (3.13)$$

Initially the cathodic product contains a variable amount of trivalent manganese in a homogeneous phase; at higher levels of discharge a new crystalline modification composed solely of $MnO.OH$ is also formed. As shown in eq.(3.5), these products may react further with zinc ions to form mixed oxides. The composition of the cathodic discharge product is thus seen to vary in a very complex manner as discharge proceeds, and this is responsible for the characteristic fall in OCV with percentage of service life expended (see, for example, Figs 3.12 and 3.15).

In addition, the slow diffusion of protons within the solid MnO_2 results in serious polarization of the cell, especially at high currents (Fig. 3.4a). If the cell is allowed to 'rest' for a period while no current is drawn, the cell voltage slowly recovers, as proton diffusion dissipates the high Mn (III) concentration near the surface of the MnO_2 grains and the composition of the solid returns to uniformity (Fig. 3.4b). (The recuperation of the cell voltage is also aided by the dispersion of the pH gradient across the cell which also depresses the OCV.)

Because of the nature of the cathodic discharge, it can be appreciated that the Leclanché cell must have severe limitations in high current operations. On the other hand, in applications requiring intermittent use, the full beneficial effects of the recuperation process may be experienced. Fig. 3.5 shows the intermittent discharge curve of a C-size cell subjected to a standard 'heavy industrial flashlight test' (4 Ω load, 4 minute discharge every 15 minutes over an 8 hour period, repeated daily): the voltage

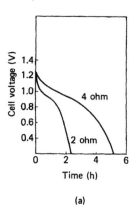

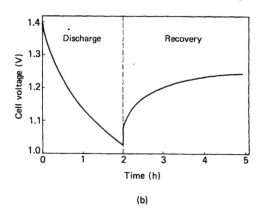

(a)

(b)

Fig. 3.4 (a) Continuous discharge curves of a D-size Leclanché cell through 2 Ω and 4 Ω loads. (b) Recovery of cell voltage after discharge through a 4 Ω load for 2 hours

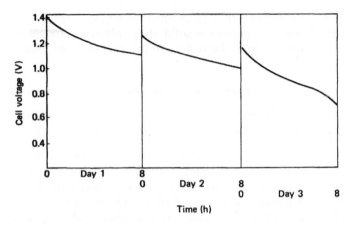

Fig. 3.5 Intermittent discharge curve of a C-size Leclanché cell through a 4 Ω load using a HIF (heavy industrial flashlight) test schedule: 4 min discharge every 15 min over an 8 hour period, repeated daily

recovery during the rest periods is clearly seen. The rate of the recuperation process is affected by the nature of the MnO_2 used.

Anode

In modern cells the anode is fabricated from a zinc alloy sheet synthesized from 'battery grade' zinc (99.99%) containing small quantities of Pb and Cd to give satisfactory mechanical properties for drawing, extrusion, etc. New forming techniques have allowed the heavy metal content to fall from about 1% by weight to under 0.1% for Cd and 0.3% for Pb. Environmental concern over disposal is likely to result in the elimination of Cd and its replacement by Mn. The potential of the zinc electrode with respect to a saturated calomel electrode is shown in Fig. 3.6 as a function of $ZnCl_2$ concentration. In the presence of NH_4Cl the potential is shifted to more negative values due to the formation of chlorozinc complexes (and amminozinc complexes at higher pH), with the consequent reduction in the activity of free zinc ions.

Electrode polarization of the anode is less severe than that for the cathode and is mainly concentration polarization resulting from accumulation of zinc chloride near the electrode surface. During rest periods after discharge, diffusion of $ZnCl_2$ into the bulk of the electrolyte reduces the anode polarization and this also contributes to the recovery of the cell voltage. Polarization behaviour of the cathode and anode of a D-size Leclanché cell is shown in Fig. 3.7.

Shelf reactions

Shelf reactions are defined as chemical processes occurring in batteries during storage before they have been used. Similar processes also occur

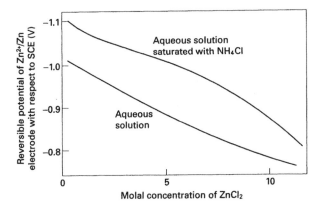

Fig. 3.6 Reversible potential of a zinc electrode with respect to a saturated calomel electrode (SCE) as a function of zinc chloride concentration

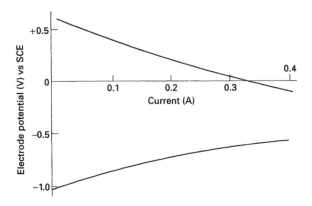

Fig. 3.7 Polarization curves for anode and cathode of a D-size Leclanché cell: current increased by steps of 25 mA once per minute

during rest periods between discharges, but where changes in electrolyte composition have occurred as a result of discharge (as happens in the case of the Leclanché cell) the nature and rates of these reactions may be greatly changed. The most significant shelf reactions which lead to the deterioration of Leclanché cells are corrosion processes at the anode. Three main mechanisms must be considered. First, if oxygen can reach the zinc/solution interface, the reaction

$$Zn(s) + 2NH_4^+(aq) + 2Cl^-(aq) + \tfrac{1}{2}O_2(g) \rightarrow$$
$$Zn(NH_3)_2Cl_2(s) + H_2O(l) \tag{3.14}$$

becomes possible. Effective sealing of the cell is therefore necessary, both to exclude oxygen and to prevent evaporation of water from the electrolyte. In the absence of oxygen, a similar corrosion reaction takes place with the

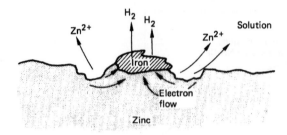

Fig. 3.8 Mechanism of zinc dissolution by the formation of a local corrosion couple

evolution of hydrogen – despite the reasonably high hydrogen overvoltage of zinc:

$$Zn(s) + 2NH_4^+(aq) + 2Cl^-(aq) \rightarrow Zn(NH_3)_2Cl_2(s) + H_2(g) \quad (3.15)$$

This reaction is greatly enhanced if the third corrosion process is widespread, namely displacement reactions which occur when electropositive impurities such as Ni^{2+}, Fe^{2+}, Cu^{2+} etc. are present in the electrolyte, e.g.

$$Zn(s) + Fe^{2+}(aq) \rightarrow Zn^{2+}(aq) + Fe(s) \quad (3.16)$$

Such displacements result in the formation of local couples where hydrogen evolution and zinc dissolution can occur at a greatly accelerated rate (Fig. 3.8). (Where the concentration of such impurities is very high, displacement reactions have been known to produce such extensive dendritic growths that the cells have become internally short-circuited.)

The corrosion reactions are strongly temperature-dependent and are also affected by the detailed morphology of the zinc and the composition of the electrolyte. Inhibition of corrosion may be achieved by a number of techniques. Until recently the principal technique was to amalgamate the zinc surface by adding a small quantity of a soluble mercury salt to the electrolyte. This increased the hydrogen overvoltage of the zinc and also allowed dissolution of small quantities of electropositive metals, thus preventing the formation of local couples. Amalgamation also influences the morphology of zinc, producing a smoother surface and reducing local corrosion. It also helps to stabilize the OCV. Again, however, environmental concerns have led to the reduction and now total elimination of mercury from Leclanché cells produced by the majority of manufacturers. It has therefore become necessary to develop other methods to prevent corrosion in Leclanché and other zinc-based cells. These have included changes in manufacturing procedure to reduce air contamination, improvements in the purity of the electrolyte and MnO_2, and the use of corrosion inhibitors. The latter are commonly chromates or dichromates which form zinc oxide films, surface active or surface-modifying agents and chelating agents.

Shelf reactions at the cathode are of minor importance, although at high temperatures some oxidation of carbon to CO_2 has been reported. Some

MnO_2 may also be lost due to reaction with the gelling agents in the electrolyte.

Water loss from the electrolyte by evaporation or reaction to form hydrates may also take place.

Leakage

A long-standing problem of Leclanché dry cells is their propensity to leak after heavy discharge. The origin of this phenomenon may be traced to the formation of insoluble complexes of zinc (involving chloride, hydroxide and ammonia) in the region between anode and cathode. The establishment of an insoluble layer drastically reduces diffusion between the two electrode layers, and in particular prevents the transport of basic species from the catholyte. The free zinc ion concentration is thus allowed to rise in the anolyte. This in turn causes the pH to fall rapidly, and the increasingly acid solution has a greater tendency to corrode the zinc, with consequent evolution of hydrogen. At the same time, the low pH hydrolyses the starch used to immobilize the electrolyte in some separator designs and produces a highly viscous phase which tends to trap the hydrogen. The pressure thus developed leads to electrolyte being forced past seals, etc. to the exterior of the cell.

Hydrogen may also be evolved at the cathode if a cell is left connected to a load after the MnO_2 has all been consumed. Such evolution will continue so long as some zinc remains connected to the load.

Very many different methods have been proposed for the reduction or elimination of the leakage problem. Most practical improvements have arisen from better design of seals, replacement of thermoplastic seals by injection-moulded seals, use of external steel containers with 'dead volume' to collect any exudate, provision of venting mechanisms, fuses, etc. An important advance was made in 1971 when a cell employing a smaller cathode volume and thicker electrolyte layer than normal was patented. In this cell, the complexed zinc precipitate was less coherent and did not interfere significantly with transport between the anode and cathode regions. No leakage was reported even when such cells were short-circuited for 24 hours. The replacement of paste/gel separators by methyl-cellulose-coated paper separators is also considered to improve the leakage problem. Many manufacturers wrap the can and seal in shrink polymer envelopes.

Typical power sources

Leclanché cells are manufactured in a range of sizes, from 11.3 mm diameter × 3.3 mm high button cells to 66.7 mm diameter × 166 mm high alarm cells. Combination of cells in series and parallel arrangements gives a wide variety of batteries with different capacities and voltages ranging from 1.5 to 510 V. The basic unit is generally constructed in

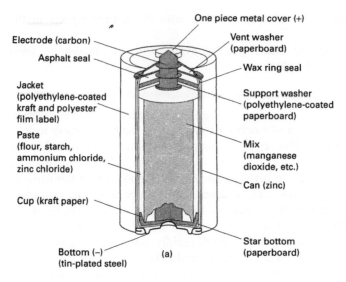

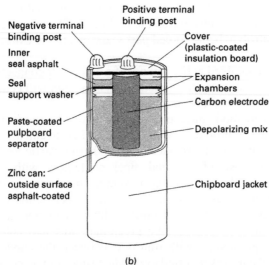

Fig. 3.9 (a) Cross-section of a standard D-size Leclanché cell. (b) Cross-section of a high capacity Leclanché alarm cell. (By courtesy of Union Carbide.)

one of two forms, namely the cylindrical cell, used alone or in groups, or the flat cell, used only in the production of multicell batteries.

The cross-sections of two typical cylindrical cells are shown in Fig. 3.9. In Fig. 3.9(a) a D-size unit used for flashlights and similar applications is shown; a large capacity alarm cell is shown in Fig. 3.9(b). The electrolyte or 'paste' separator in Fig. 3.9(a) is a relatively thin layer of electrolyte solution immobilized in a gel or microporous separator. Different manufacturers favour different forms of separator. These range from gelled electrolytes of varying thickness, using either natural cereals or synthetic polyvinyl or cellulose alkyl ethers, pulp board liners soaked in electrolyte solution (as in Fig. 3.9b) and special papers coated on either side with films of gelling agent particularly suitable for anolyte or catholyte. Similarly, different approaches are made to sealing, dealing with leakage, external jacket material, etc., according to the preference of the manufacturer.

An alternative design for cylindrical cells is the so-called 'inside-out' configuration, in which the anode, in the form of zinc sheet vanes, is centrally placed and surrounded by the cathode mix. The principal advantage of this design is its high leakage resistance which results from the fact that the cell case is no longer fabricated from zinc.

The second basic design, the flat cell, is illustrated in Fig. 3.10. In Fig. 3.10(a) a section of the unit cell is shown; Fig. 3.10(b) shows a 9 V series assembly of six cells. Batteries of this type are based on 'duplex' or 'bipolar' electrodes or carbon-coated zinc which act as the cathode current collector for one cell and the anode for the adjacent cell. The electrolyte/separator usually consists of one or two layers of different electrolyte-impregnated papers which have been treated with suitable gelling agents. The cathode mix of MnO_2, acetylene black and electrolyte is formed into a flat cake and each cell is held together by a plastic band, as shown. The group of six cells are sealed with a wax coating and assembled within a metal jacket. No provision of expansion volume is found necessary with this form of construction, and a relatively high energy to volume ratio is realized.

A special type of flat battery has been developed for use in automatic or self-developing cameras by Rayovac: an exploded view of this battery system is shown in Fig. 3.11(a). Both cathode and anode mixes contain a water-based latex binder and are applied or 'painted' onto a conductive sheet and then dried. A thin synthetic netting just over 0.1 mm thick is used to absorb the gelled electrolyte, to separate successive duplex electrodes, and to act as a support for the hot-melting sealing compound applied around its perimeter. This advanced technology battery containing four cells has a cross-sectional area of 7.0 cm × 8.5 cm and a thickness of less than 3 mm. It weighs about 19 g. Another slim Leclanché cell is the 'paper battery' produced by Matsushita (Fig. 3.1 and Fig. 3.11b). This 1.5 V cell is manufactured in a number of different shapes and has a maximum thickness of 0.8 mm. A circular cell with a diameter of 38 mm weighs 1.5 g.

Performance

The practical capacity of a Leclanché cell does not have a fixed value since it varies according to the pattern and conditions of discharge – to a much greater extent than for most other cells. Different forms of this type of cell are designed for particular purposes, and as pointed out in Chapter 2, test procedures imitate as far as possible the discharge pattern of the applica-

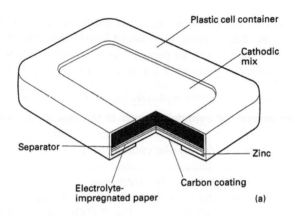

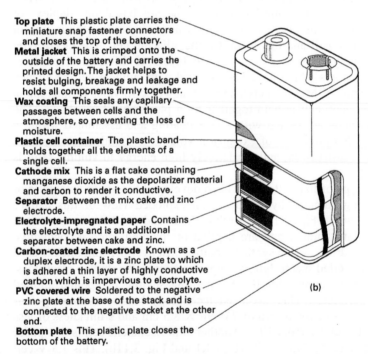

Top plate This plastic plate carries the miniature snap fastener connectors and closes the top of the battery.

Metal jacket This is crimped onto the outside of the battery and carries the printed design. The jacket helps to resist bulging, breakage and leakage and holds all components firmly together.

Wax coating This seals any capillary passages between cells and the atmosphere, so preventing the loss of moisture.

Plastic cell container The plastic band holds together all the elements of a single cell.

Cathode mix This is a flat cake containing manganese dioxide as the depolarizer material and carbon to render it conductive.

Separator Between the mix cake and zinc electrode.

Electrolyte-impregnated paper Contains the electrolyte and is an additional separator between cake and zinc.

Carbon-coated zinc electrode Known as a duplex electrode, it is a zinc plate to which is adhered a thin layer of highly conductive carbon which is impervious to electrolyte.

PVC covered wire Soldered to the negative zinc plate at the base of the stack and is connected to the negative socket at the other end.

Bottom plate This plastic plate closes the bottom of the battery.

Fig. 3.10 (a) Cross-section of a flat Leclanché cell. (b) 9 V battery of six flat Leclanché cells. (By courtesy of Berec.)

tion. Leclanché cells are sometimes classified as 'general purpose' (GP) or 'heavy duty' (HD). The former use a high proportion of NMD, a mainly NH_4Cl-based electrolyte and an inexpensive starch paste separator. Such cells have the lowest cost of all primaries but are suitable for intermittent low-rate discharges. HD cells which use a significant proportion of EMD or CMD, a higher $ZnCl_2$: NH_4Cl ratio and a paper separator, are a relatively inexpensive solution for many medium-rate continuous discharges or

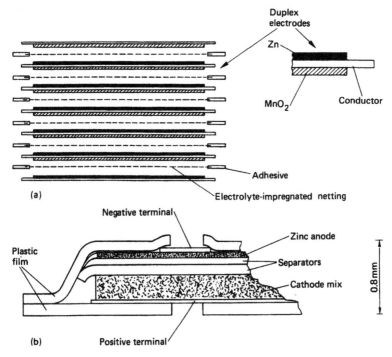

·**Fig. 3.11** (a) Cross-section of a 3 mm thick Leclanché battery used in automatic cameras, after a drawing by Rayovac. A detail of the construction of the duplex electrodes is shown. (b) Cross-section of the Matsushita 'paper battery'

heavy intermittent service. Thus, in comparing performance of Leclanché cells, it is important always to compare like with like.

The factors which affect the electrical output of a Leclanché cell may be divided into two groups: (1) cell-dependent, e.g. construction, composition and size, and (2) user-dependent, e.g. discharge rate, cut-off voltage, operating schedule, operating temperature, storage conditions, etc.

The construction of a cell involves a number of variables such as the relative amount and nature of the cathodic mix, type of separator used, etc. Fig. 3.12 illustrates the difference in service life between a standard GP cell and a HD cell which has been designed for high power applications when both are subjected to a heavy discharge (30 minutes once a day through a 2 Ω load). For a similar construction and composition, and a fixed duty schedule, higher capacities are obviously obtained by using cells of larger size.

The effect of discharge rate is shown in Figs 3.13 and 3.14 where closed circuit voltage characteristics and service life of D-size cells are shown as a function of current drain. The fact that service life increases as current density decreases suggests that for this type of system it is always best to use as large a cell as possible. It has been estimated that over a wide range of current densities, the service life is tripled by halving the current density.

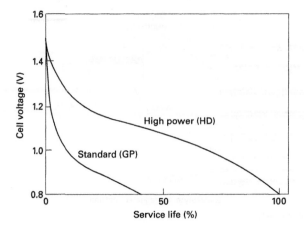

Fig. 3.12 Difference in service life of standard (GP) and 'high power' (HD) Leclanché D-size cells discharged for 30 min/day through 2 Ω loads

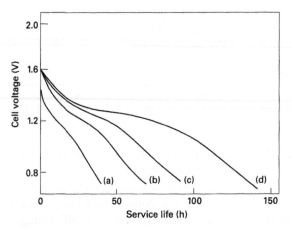

Fig. 3.13 Effect of discharge rate on service life of D-size Leclanché cells discharged at 2 hours/day: (a) initial drain 150 mA; (b) initial drain 100 mA; (c) initial drain 75 mA; (d) initial drain 50 mA

The service schedule is of critical importance in relation to the recuperation reaction. Unless the current drain is very low, Leclanché cells give a much better performance when used on an intermittent basis. The effect of rest periods was shown earlier in Fig. 3.5. In Fig. 3.15 the effect on the service life of D-size cells of two different operating schedules is illustrated. A schematic three-dimensional representation of capacity as a function of current drain and operating programme based on Union Carbide technical data is given in Fig. 3.16.

Other factors which affect the electrical output are the operating temperature and the storage conditions. Cells are generally tested at 21°C (70°F). Higher temperatures increase the energy output but reduce the shelf

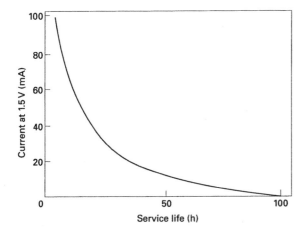

Fig. 3.14 Effect on service life of initial current drain. D-size standard Leclanché cell discharged at 4 hours/day or to 0.9 V cut-off

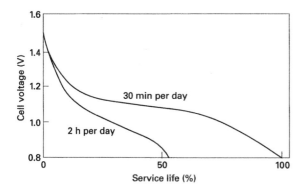

Fig. 3.15 Effect of operating schedule on service life of a standard D-size Leclanché cell discharged through a 5 Ω load

life. The consequences of altering the operating temperature on short- and long-term discharge are shown in Fig. 3.17. For long-term discharge (6 months) at temperatures over 30°C, the detrimental effects of accelerated shelf reactions are dominant. In Table 3.1 the normalized capacity of D-size cells, discharged continuously through a load resistance of 2.25 Ω to a cut-off voltage of 0.9 V, is given as a function of temperature. As is clearly seen, the low temperature behaviour of normal Leclanché cells is very poor. Special Leclanché cells with altered electrolytes have been used in the past for low temperature applications, but in current practice such cells are replaced by lithium, zinc chloride or zinc/alkaline MnO_2 cells which have greatly superior low temperature behaviour.

Advances in Leclanché dry cell performance have been continuous since the time of Geissner. Between 1900 and 1960, the specific capacity of cells

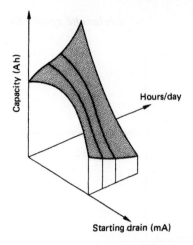

Fig. 3.16 Capacity of D-size standard Leclanché cells as a function of duty cycle and initial current drain. (By courtesy of Union Carbide.)

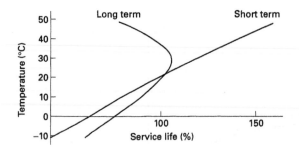

Fig. 3.17 Effect of temperature on the service life of a 9 V Leclanché flat battery subjected to short-term (~1 month) or long-term (~6 months) discharge. (By courtesy of Berec.)

Table 3.1 Effect of temperature on the capacity of D-size Leclanché and $ZnCl_2$ cells when discharged continuously through 2.25 Ω to a cut-off voltage of 0.9 V

Temperature (°C)	Normalized capacity	
	Leclanché	$ZnCl_2$
37.8	1.40	1.15
26.7	1.10	1.05
21.1	1.00	1.00
15.6	0.90	0.95
4.4	0.70	0.85
−6.7	0.45	0.70
−17.8	0.25	0.45

doubled roughly every 20 years. Between 1960 and 1980 a further increase in capacity of over 50% has been achieved. Despite the rapid evolution of new forms of primary battery in recent years, the hold of the Leclanché cell on the market remains very strong for those applications for which its relatively poor discharge characteristics are adequate. The reasons for this are not hard to find: the materials used in its construction are readily available and relatively cheap, while fabrication processes are comparatively straightforward. There is, however, now a strong trend, particularly in the USA and Europe, to replace Leclanché technology with the closely related zinc chloride cells or, for more severe service schedules, the alkaline manganese cells which are now considered below.

3.3   Zinc chloride cells

The so-called 'zinc chloride cells' are basically Leclanché cells in which the ammonium chloride has been completely, or almost completely, replaced by $ZnCl_2$. (Up to 1% of NH_4Cl is reported to improve high rate performance.) The resulting cells have a better service capacity at high current drain, at low temperatures and on continuous discharge.

The cell may be written as

$$Zn(s)|ZnCl_2(aq)|MnO_2(s),C(s)$$

and the OCV is again about 1.5 V. The accepted overall cell reaction is

$$4Zn(s) + ZnCl_2(aq) + 8MnO_2(s) + 8H_2O(1) \rightarrow$$
$$8MnO.OH(s) + ZnCl_2.4Zn(OH)_2(s) \qquad (3.17)$$

It should be noted that water is consumed in the cell reaction so that there is a tendency for the cell to dry out during discharge unless sealing arrangements are adequate. On the other hand, $ZnCl_2$ cells tend to have much greater resistance to electrolyte leakage as a result. The anode and cathode are similar to those in the Leclanché cell, although higher quality MnO_2 and a higher percentage of acetylene black are normally used. The advantages of this system arise from the properties of the electrolyte where higher rates of diffusion are possible, since there is less tendency for the electrolyte layers near the electrode surfaces to be blocked by insoluble products. The range of operating temperatures is much wider (see Table 3.1), so that this system can form a satisfactory low temperature power supply. Further, higher current densities may be obtained without unacceptable polarization; for this reason some manufacturers refer to $ZnCl_2$ cells as 'heavy duty' batteries.

Fig. 3.18 shows a section of a commercial D-size $ZnCl_2$ cell. A more sophisticated sealing system is generally employed since prevention of leakage is more important with the more acid electrolyte and since drying out and oxygen ingress must be avoided. $ZnCl_2$ cells always use a thin paper separator and generally have resealable gas vents.

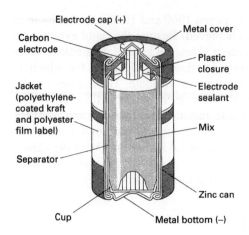

Fig. 3.18 Cross-section of a D-size zinc chloride cell. (By courtesy of Union Carbide.)

The improved performance of the $ZnCl_2$ cell is offset by the higher cost incurred through using a better quality cathode mix and more complex fabrication due to the requirement of more reliable seals. Since the improvement in performance out-matches the price differential between Leclanché and zinc chloride cells, there is an increasing tendency to phase out the classical Leclanché formulation in favour of the latter. The best zinc chloride cells, sometimes referred to as 'extra heavy duty' (EHD), have at least twice the service life of even the best Leclanché cells.

In recent years, there have been further improvements in the performance of zinc chloride cells. These have been brought about principally by use of laser welding to increase the internal active volume, and by improvements to materials, especially by synthesizing acetylene black with optimized surface area and water absorption characteristics.

3.4   Alkaline manganese cells

This type of cell is another variant on the basic Leclanché cell. In this case, the electrolyte is a concentrated aqueous solution of potassium hydroxide (about 30%), partly converted to potassium zincate by the addition of zinc oxide. The main advantage of alkaline manganese cells over Leclanché cells is their relatively constant capacity over a wide range of current drains and under severe service schedule conditions. Another feature of this system is that it can be the basis of a secondary battery system. The cell reaction may be written formally as

$$2Zn(s) + 2MnO_2(s) + H_2O(1) \rightarrow 2MnO.OH(s) + 2ZnO(s) \quad (3.18)$$

but is in practice much more complex than this due to further reduction of the manganese, as discussed below, and the formation of various soluble zincate species. The OCV is 1.55 V at room temperature. A wet cell based on this system was reported in 1882, but the first commercial dry cell was not available until 1949 and major commercialization did not occur until the 1960s.

Largely because of the high conductivity of the electrolyte in comparison with those in the Leclanché or zinc chloride cells, the alkaline manganese battery has a much superior performance, especially at high rates. While the production of the former cells is levelling off in some countries (and in the USA is decreasing), the number of alkaline manganese cells is rapidly increasing. In 1994 the 'alkaline ratio', i.e. the ratio of alkaline manganese cells to all aqueous zinc primaries, was reported as:

- USA – 72%

- Europe – 52%

- Japan – 29%

The alkaline manganese dioxide cell is most widely available either as standard sized cylindrical cells with capacities ranging from 0.6 to 22 Ah or as button cells. Batteries having a wide variety of capacities and voltages are also readily available. These are all interchangeable with Leclanché and zinc chloride cells.

Electrolyte and separator

A variable quantity of ZnO is added to the concentrated KOH solution, depending on the system characteristics required. ZnO can also act as a gassing suppressor. The electrolyte is immobilized generally using carboxy-methylcellulose, and a non-woven fabric separator made of natural or synthetic fibres resistant to the high pH is placed between the electrodes.

Cathode

The cathode mix is a compressed mixture of electrolytic MnO_2 (EMD) and synthetic graphite or acetylene black to provide electronic conductivity, in a ratio of 4–5: 1, wetted with electrolyte. The cathode current collector is generally the external steel can, which may be nickel-plated or coated with conductive carbon. Reduction of MnO_2 in alkaline conditions is a complex process and follows a number of steps which can be written formally as

$$MnO_2 \rightarrow MnO_{1.5}$$
$$MnO_{1.5} \rightarrow MnO_{1.33}$$
$$MnO_{1.33} \rightarrow MnO$$

The last two stages are only possible at very low current drain. A schematic diagram of cell voltage as a function of the degree of reduction of MnO_2 is shown in Fig. 3.19. Since MnO.OH forms a solid solution with MnO_2, the first stage gives rise to the characteristic sloping discharge behaviour.

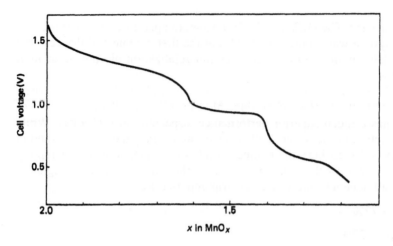

Fig. 3.19 Cell voltage of an alkaline manganese cell as a function of the degree of reduction of MnO_2

Provided that the reduction does not exceed a level equivalent to $MnO_{1.33}$ the reaction can be reversed and the cathode recharged. In practice this means limiting the discharge at 0.9 V. The rechargeable alkaline manganese (RAM) cell is discussed in Chapter 6.

Note that the cathode components must all have very low levels of electropositive metal impurities in order to minimize gassing at the anode were such impurities to migrate to its surface.

Anode

The anode is a hollow cylinder of powdered zinc set in a carboxymethylcellulose, polyacrylate or other polymer-based gel. Zinc powder with median particle diameters in the range 150–250 μm is produced by interaction of a thin stream of molten metal with jets of compressed air. Gassing may be reduced by alloying the zinc with small quantities of aluminium, bismuth or calcium. As in the case of Leclanché cells, manufacturers use a range of organic additives to reduce zinc corrosion now that mercury amalgamation is no longer acceptable for environmental reasons. The ratio of zinc to gelled electrolyte is controlled to ensure electronic contact between particles while avoiding complete blockage of ionic pathways by ZnO. The current collector is usually made of brass strip or pins in the cylindrical cells, and is the stainless steel lid of the button cells. The anode collector surfaces become coated with zinc on cell assembly.

Typical power sources

A cutaway section of a cylindrical primary cell is shown in Fig. 3.20. The steel construction and intricately engineered sealing assembly reduce the

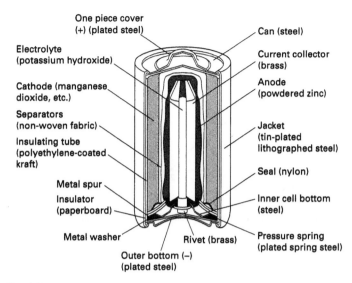

One piece cover
(+) (plated steel)

Electrolyte
(potassium hydroxide)

Cathode (manganese
dioxide, etc.)

Separators
(non-woven fabric)

Insulating tube
(polyethylene-coated
kraft)

Metal spur

Insulator
(paperboard)

Metal washer

Outer bottom (−)
(plated steel)

Rivet (brass)

Can (steel)

Current collector
(brass)

Anode
(powdered zinc)

Jacket
(tin-plated
lithographed steel)

Seal (nylon)

Inner cell bottom
(steel)

Pressure spring
(plated spring steel)

Fig. 3.20 Cross-section of a D-size alkaline manganese primary cell. (By courtesy of Union Carbide.)

risk of leakage of the highly caustic electrolyte or of uncontrolled release of internal pressure. In more recent designs, a larger usable volume is achieved by using a thin plastic jacket or foil label. Further increases in capacity have been made possible by use of butt-seam welding and lower profile seals. More sophisticated safety devices are often incorporated to deal with gas generation caused in series-connected cells where there is either a reverse connected cell or a mixture of new and partially discharged cells.

Performance

The discharge curve for a D-size alkaline manganese primary cell with an initial current drain of 500 mA is shown in Fig. 3.21. Compared with even the best Leclanché cell of this size, the alkaline manganese cell has a capacity advantage of at least a factor of 4 in high current continuous discharge applications. The shelf life of these cells is also good: after a 4 year storage at 21°C, they retain more than 80% of their initial capacity. At low temperatures, especially at low current drain, their performance is as good as that of the $ZnCl_2$ cell. However, because of their higher cost (as with the $ZnCl_2$ cell, due to the use of high quality materials and a much more sophisticated construction), there is an economic advantage in using them to replace Leclanché cells only when heavy duty service is required. Thus for intermittent use, it is not expedient to replace D-size cells below 300 mA current drain.

It is perhaps useful at this stage to compare the performances of all the Zn/MnO_2-based primary systems. In Fig. 3.22 the discharge curves of four

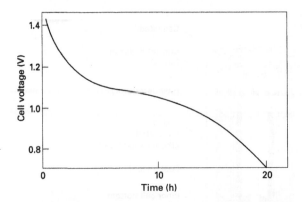

Fig. 3.21 Discharge curve of a D-size alkaline manganese primary cell with an initial current drain of 500 mA

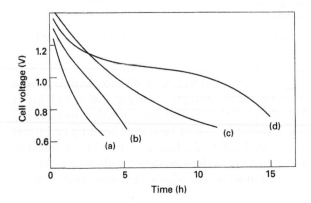

Fig. 3.22 Comparison of the performances of Zn–MnO$_2$ primary systems under 2.25 Ω continuous test: (a) standard Leclanché cell based on natural ore; (b) 'high power' Leclanché cell based on electrolytic MnO$_2$; (c) zinc chloride cell; (d) alkaline manganese cell

D-size cells on 2.25 Ω continuous test are reported. Cell (a) is a standard Leclanché cell using a natural ore; cell (b) is a HD Leclanché with electrolyte MnO$_2$; cell (c) is a zinc chloride cell; and cell (d) is an alkaline manganese primary unit. The differences at this current drain are striking: the discharge capacities with a 0.9 V cut-off are in the ratio 0.12:0.24: 0.55:1.00 for the four types. However, when less severe tests are considered, the disparities are less pronounced. Thus for the light industrial flashlight (LIF) test, the ratios are 0.40:0.61:0.96:1.00.

Applications

Cylindrical cells are widely used in devices such as automatic cameras, toys, headphone stereos, radios, liquid crystal TVs and other portable AV and communication equipment. They are also used in applications such as emergency lighting. Button cells are used where the lower cost compensates for the lower energy density of competing systems and where the sloping discharge curve is acceptable.

3.5 Aluminium- and magnesium-based Leclanché cells

Replacement of zinc anodes by aluminium or magnesium seems an attractive proposition because of their significantly higher specific capacities (2.98 Ah/g for Al and 2.20 Ah/g for Mg compared with 0.82 Ah/g for Zn). In addition, both metals have higher standard potentials than zinc so that a higher cell voltage and energy density can be anticipated. Two problems have delayed the development of practical systems and limited commercial exploitation to a few types of magnesium-based cell: the first is the greatly increased corrosion rate, and the second is the presence of an oxide film which limits anode corrosion but is responsible for a 'voltage delay' on discharge.

Passivating films

The effect of passivating films on aluminium and magnesium has been the subject of much research. By incorporating chromate/dichromate mixtures and other substances in the electrolyte, a coherent insoluble oxide film is formed which effectively inhibits further corrosion. Sealed cells with aluminium or magnesium anodes may therefore be successfully stored for several years, even at high temperatures. However, once current has been drawn from the cell, the film is broken down and rapid attack on the metal follows due to reactions such as

$$Mg(s) + 2H_2O(1) \rightarrow Mg(OH)_2(s) + H_2(g) \qquad (3.19)$$

until finally a passive film is reformed. Such corrosion processes reduce the anode capacity, remove water or hydrogen ions from the electrolyte, and produce hydrogen gas which requires venting. The reaction is also highly exothermic, so that heat dissipation is a problem at high discharge rates. Note that the corrosion reaction occurs in parallel to the cell reaction so that at best only one-half to two-thirds of the magnesium is used to generate current.

The presence of passivating films reduces the cell voltage below the anticipated thermodynamic value calculated assuming a simple metal/metal ion process at the anode. More important, however, is the fact that the films are responsible for a time-lag between the point at which a current drain is initiated and the point at which the cell reaches its operating voltage. An example of this voltage delay is shown in Fig. 3.23 where

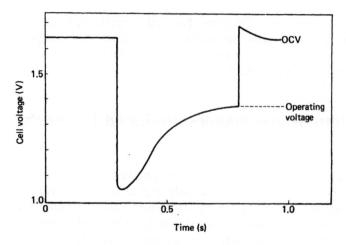

Fig. 3.23 Voltage delay in an aluminium-based D-size Leclanché cell subjected to a 500 mA pulse for 0.5 s

the cell voltage of an aluminium-based D-size Leclanché cell is plotted as a function of time during the passage of a 0.5 s, 500 mA current pulse. The initial severe polarization is due to a combination of the low effective exchange current caused by the presence of the film and the ohmic resistance of the film itself.

As the film is disrupted by the passage of current, the cell voltage rises to a steady operating value. The length of the voltage delay is dependent on the anode material, the electrolyte composition and the discharge history, and may extend from a few hundred milliseconds to 20 s or more. The 'overshoot' in cell voltage when the cell is returned to open circuit corresponds to the period in which a new film is formed; this may be contrasted with the behaviour of lithium in organic solvents (Fig. 4.6). Further consideration is given to voltage delay phenomena in Chapter 4 where lithium anodes are discussed.

Cell constitution

Aluminium-based cells are generally formulated with $AlCl_3$ or $CrCl_3$ solutions as electrolyte. The optimum electrolytes for magnesium-based cells are $MgBr_2$ or $Mg(ClO_4)_2$, buffered with $Mg(OH)_2$ to a pH of approximately 8.5. Chromate inhibitors are always added: the exact choice of inhibitor affects the extent of voltage delay phenomena. Resistance to corrosion and shorter voltage delay can also be obtained by using suitable alloys for the anodes. In the case of magnesium, addition of zinc ($\approx 1\%$) reduces the delay, and aluminium ($\approx 2\%$) optimizes the current efficiency with respect to the corrosion reaction. Similarly for aluminium, a number of alloy compositions and heat treatments have been recommended. The cathode usually resembles closely the MnO_2 system of the standard

Leclanché cell. Experimental cells with silver, mercuric and other oxides have been developed but have not been exploited commercially.

Performance

A characteristic feature of cells having aluminium or magnesium anodes is their higher working voltage in comparison with their zinc analogues: thus, the magnesium-based Leclanché cell has an OCV of approximately 1.9 V. The capacity of these cells is, however, very variable, being dependent on the extent of the corrosion reaction, which is in turn a function of the discharge regime. For intermittent service, practical capacities as low as 40% are common, whereas high rate discharge may furnish 70% of the theoretical capacity, and so give a specific capacity of over twice that of a conventional Leclanché cell.

A number of cylindrical and flat magnesium-based cells have been developed on a commercial scale, mainly for military applications where high discharge currents and low unit weight are important. However, for most of these applications, magnesium batteries have now been replaced by various lithium/organic systems. There are no commercial aluminium-based Leclanché cells. Magnesium and aluminium are both exploited as anodes in metal–air cells which are considered below.

3.6 Zinc–mercuric oxide, cadmium–mercuric oxide, zinc–silver oxide and related systems

'Miniature' or 'button' cells are cylindrical in form and have a height of less than 5 mm. The market for these cells has exploded over the past 20 years due to the development of electric watches and other miniature electronic devices and current production is over 10^9 units per annum. The various systems all share two favourable features, namely high volumetric capacity, which is relatively unaffected by current drain, and good discharge characteristics, even under conditions of relatively heavy discharge. The earliest aqueous system, based on mercuric oxide and zinc, was introduced in the 1940s. This is the Ruben–Mallory or RM cell, a classic combination of electrochemical and engineering ingenuity which revolutionized the battery industry. The zinc–silver oxide system was introduced commercially by Union Carbide in 1961 shortly after the appearance of electric watches, and a number of other alkaline electrolyte button cells were developed subsequently. Nowadays such cells are being displaced by lithium primaries which have superior energy density, are in some cases less expensive, and are seen as being more environmentally friendly.

The zinc–mercuric oxide system

This system, commonly known as the 'mercury cell', is based on an amalgamated zinc anode, a concentrated aqueous potassium hydroxide electrolyte – saturated with zincate ion by zinc oxide – and a mercuric oxide/graphite cathode:

$$Zn(s)|ZnO(s)|KOH(aq)|HgO(s),C(s)$$

The anode reaction may be written as

$$Zn(s) + 2OH^-(aq) \rightarrow ZnO(s) + H_2O(l) + 2e \qquad (3.20)$$

and the cathode reaction as

$$HgO(s) + H_2O(l) + 2e \rightarrow Hg(l) + 2OH^-(aq) \qquad (3.21)$$

so that the overall cell reaction is

$$Zn(s) + HgO(s) \rightarrow ZnO(s) + Hg(l) \qquad (3.22)$$

Two important points to note are the invariance of the electrolyte solution and the constancy of the chemical potentials of reactants and products, as the discharge proceeds. One consequence of the effective non-involvement of the electrolyte is that only a very small quantity is required in a working cell. Another is a relatively constant internal resistance, leading to a flat discharge curve. The constancy of chemical potentials implies a constant OCV during the course of the discharge, as discussed in Chapter 2. The free energies of formation of $HgO(s)$ and $ZnO(s)$ are given as 58.4 and 318.2 kJ/mol, respectively, in Latimer's *Oxidation potentials*. Hence, the free energy change for the cell reaction is 259.8 kJ/mol and the cell emf is 1.347 V, which is in very satisfactory agreement with the OCV of 1.357 V of commercially produced cells. The OCV may be slightly increased by adding MnO_2 to the cathode mix.

The electrolyte is usually an approximately 40% solution of KOH saturated with zinc oxide, to which corrosion inhibitors have been added. The KOH is occasionally replaced by NaOH. Caustic soda solutions have a lower tendency to creep, but have a higher electrical resistance. The electrolyte is immobilized using felted cellulose. The most common anode is a porous compressed cylindrical pellet of amalgamated zinc powder and electrolyte (possibly gelled). An alternative configuration is the 'wound anode' usually found in secondary zinc–mercury oxide cells, which uses a spiral of corrugated zinc foil interleaved with an absorbent paper strip. (The corrugations increase the surface area and provide adequate volume for the deposition of zinc oxide during discharge.) The cathode pellet consists of mercuric oxide together with 5–10% of finely divided graphite, added to increase the electronic conductivity and to minimize the coalescence of mercury formed during discharge. The cathode always has a larger capacity than the anode. The cell is therefore 'zinc limited', i.e. in an exhausted cell there is no zinc left which might corrode and thus lead to

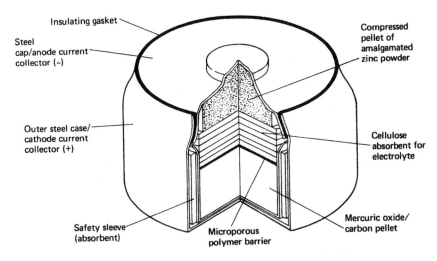

Fig. 3.24 Cross-section of a typical zinc–mercuric oxide button cell

hydrogen pressure developing in the cell. A microporous plastic barrier layer is generally placed next to the cathode pellet to prevent internal short circuits caused by the displacement of free mercury or graphite. More complex multiple-barrier layers may be necessary if the cell has been designed for low current drain operation.

The cross-section of a typical mercury button cell is shown in Fig. 3.24. The cathode and anode current collectors are the steel case and steel top, respectively. Attention is drawn to the sophisticated engineering design of this cell, which has provision for automatic venting of any pressure caused by hydrogen evolution, with any electrolyte displaced being absorbed in the safety sleeve between the inner and outer case.

Mercury cells have practical specific capacities of up to 400 Ah/dm^3 and specific energies of 550 Wh/dm^3. In addition, they have particularly flat discharge characteristics, even under conditions of continuous discharge. These are nearly independent of load over a wide range, as seen in Fig. 3.25. The flat part of the curve is known as the 'equilibrium region' and may extend to 97% of the cell capacity at low current drains (e.g. 1 mA for a 20 mm diameter button cell). Momentary short circuits do not damage the cell and voltage recovery is rapid. Storage behaviour is good, with over 90% of the initial capacity retained after 1 year. Low temperature performance is not particularly favourable, although it is improved if the wound anode configuration is used, especially if the current demands are low or intermittent.

Power sources based on the zinc–mercuric oxide system are particularly suited to a wide range of applications, mainly concerned with miniature portable electronic equipment, where a relatively constant voltage is required throughout long discharge periods. In addition, such cells are used as voltage reference standards in regulated power supplies, potentiometers, chart

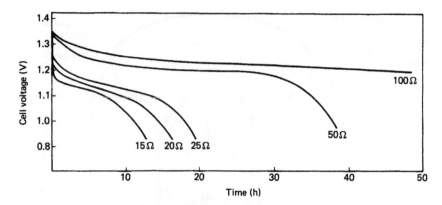

Fig. 3.25 Discharge characteristics of 1 Ah zinc–mercuric oxide button cell under continuous load at room temperature

recorders, etc. The market is shared with the more expensive zinc–silver oxide system described below.

The cadmium–mercuric oxide system

Replacing zinc with cadmium produces a cell with an OCV of 0.90 V, with characteristics very similar to those of the zinc–mercuric oxide system described above, but which is able to be stored and operated at extreme temperatures (−55 to 80°C) due to the low solubility of cadmium oxide even in concentrated KOH. Cells have been successfully operated at 180°C. Note that hydrogen generation does not occur at a cadmium anode. Because of cost and disposal problems, such cells are used only for applications where their special properties can be exploited, e.g. telemetry from internal combustion, jet or rocket engines.

The zinc–silver oxide system

The cell may be written as

$$Zn(s)|ZnO(s)|KOH(aq)|Ag_2O(s),C(s)$$

The main features of zinc–silver oxide cells are similar to those of the zinc–mercuric oxide system, except for a higher OCV and significantly increased cost. The overall cell reaction is

$$Zn(s) + Ag_2O(s) \rightarrow ZnO(s) + 2Ag(s) \qquad (3.23)$$

The emf calculated from Latimer's *Oxidation Potentials* is 1.593 V, which agrees well with the OCV of commercial cells of 1.60 V. A cutaway view of a typical silver oxide button cell is shown in Fig. 3.26: details of the sealing arrangement vary from manufacturer to manufacturer. Since Ag_2O has a very high electrical resistance, 1–5% by weight of graphite is generally added to the anode. As the cell is discharged, silver is produced

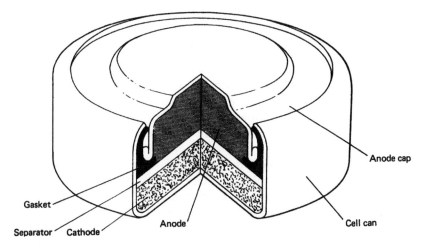

Fig. 3.26 Cut-away view of a typical zinc–silver oxide button cell. (By courtesy of Union Carbide.)

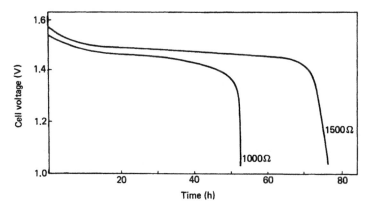

Fig. 3.27 Discharge characteristics of 75 mAh zinc–silver oxide hearing aid cell under continuous load at room temperature. (By kind permission of P. Ruetschi, Plenary lecture, ISE, Venice, 1980.)

which improves the conductivity. One manufacturer uses granular Ag_2O which is pretreated with a reducing agent to form a surface layer of silver which obviates the need for graphite. Other cells use a mixture of Ag_2O and $AgNiO_2$, which is both an electronic conductor and an active cathode material.

Separators must be resistant to the high pH and, since silver oxide is slightly soluble in strong bases, must prevent migration of silver ions to the anode. Such separators are further considered in Chapter 6, where secondary batteries based on the zinc–silver oxide system are described.

Discharge curves for a 75 mAh hearing aid battery are shown in Fig. 3.27 for two typical loads. Detailed differences in formulation are made in the production of cells for different uses. For example, electronic watches and other devices with liquid crystal diode (LCD) displays require currents of 3–10 μA, and high resistance batteries with NaOH electrolyte are suitable. On the other hand, watches with light emitting diode (LED) displays or with LCD displays, together with additional electrical illumination or alarms, require batteries with low internal resistance which can maintain a stable voltage while supplying current pulses of up to 70 mA for 1 or 2 s. Cells for such applications use KOH electrolyte. This electrolyte is also used for hearing aid batteries where continuous current drains of 1–2 mA are typical.

Another application for somewhat larger zinc–silver oxide cells is in the production of high voltage, high capacity reserve batteries for military equipment, e.g. for missile guidance systems and electrically driven torpedo motors. Such batteries are remotely activated by an electric signal which causes a diaphragm in an electrolyte reservoir to be broken, or electrolyte to be forced into the cells under pressure. The electrolyte solution may be contained in a tubular coil, in cylindrical tanks fitted with pistons or in pressure-over-liquid (POL) tanks. Missile batteries together with their electrolyte delivery systems are carefully assembled and packed in metal containers of sophisticated design and are able to sustain shock levels of up to 2000 g and other extremes of mechanical and environmental stress. They usually have overall weights in the range 1–20 kg. Their primary function is to energize hydraulic power systems for the missile guidance surfaces. Torpedo batteries must be able to deliver 300 kW for up to 10 minutes. In order to fit into the torpedo body they must be cylindrical in form, with a plate diameter determined by the internal dimensions of the torpedo casing. This requirement, coupled with the high currents drawn ($\approx$ 500 A), necessitates the use of a bipolar stack construction. Such batteries can weigh up to 400 kg and have practical energy densities of over 80 Wh/kg.

Zinc–silver oxide reserve batteries have application as power sources for various systems in manned and unmanned space vehicles.

Related systems

A number of primary cells using zinc and aqueous KOH electrolyte but with alternative cathode materials have been developed. Divalent silver oxide, AgO, has a higher energy density and nearly twice the capacity of Ag_2O, but it is metastable and tends to decompose by evolving oxygen. It also has a two step discharge (1.8 and 1.6 V) corresponding to

$$AgO(s) + \tfrac{1}{2}H_2O(l) + e \rightarrow \tfrac{1}{2}Ag_2O(s) + OH^-(aq) \qquad (3.24)$$

and

$$AgO(s) + H_2O(l) + 2e \rightarrow 2Ag(s) + 2OH^-(aq) \qquad (3.25)$$

respectively. Research into stabilizing agents has made it possible to use a mixture of the two oxides in miniature cells which have a higher capacity

Table 3.2 Practical energy density of miniature alkali button cells (diameter, 11.6 mm; height, 4.2 mm) with zinc anodes at a discharge rate of 5–10 μA

Cathode material	Midlife voltage	Volumetric energy density (Wh/cm^3)
MnO_2	1.30	0.23
$NiO(OH)$	1.55	0.23
Ag_2O	1.55	0.45
CuO	0.90	0.50
HgO	1.35	0.53
AgO	1.55	0.60
Air	1.25	0.95

than the standard Ag_2O cells. Where the two step discharge is unacceptable, various ideas have been exploited to eliminate it. For example, Ag_2O particles may be treated by chemical reducing agents to form first a thin layer of AgO and then a surface layer of Ag. The potential of the cathode is thus reduced to 1.6 V, but the capacity corresponds to the Ag_2O. Another method involves the formation of $Ag_5Pb_2O_6$ or $AgBiO_3$.

Other cathodes which have been studied include CuO and NiO(OH). The energy densities of a number of systems (including 'alkaline manganese' and 'air-depolarized' zinc cells for comparison) are given in Table 3.2.

The primary objective of miniature battery design is to maximize the energy density in a small container. A compromise must be reached, however, since volumetric energy density decreases as cell volume decreases and the 'dead volume' due to containers, seals, etc. becomes increasingly significant. A plot of energy density as a function of total volume is given in Fig. 3.28 for the zinc–mercuric oxide and zinc–silver oxide systems.

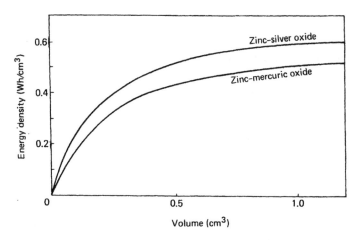

Fig. 3.28 Energy density as a function of total volume for the zinc–mercuric oxide and the zinc–silver oxide systems. (By kind permission of P. Ruetschi, Leclanché SA, Yverdon, Switzerland.)

3.7   Metal–air batteries

A number of cells have been developed which make use of the oxygen of the air as the cathodic reagant. These so-called 'air-depolarized cells' are examples of hybrid cells, which are discussed more fully in Chapter 9. Many recent advances in metal–air batteries can be attributed to the research carried out in the 1960s on high current density air electrodes for ambient hydrogen/oxygen fuel cells using aqueous electrolytes.

The most obvious advantages of the oxygen cathode are that it has low weight and infinite capacity. Consequently, prototype D-size cells based on the zinc–air system have been shown to have twice the overall practical capacity of zinc–mercuric oxide cells (and 10 times that of a standard Leclanché cell) when subjected to a continuous current drain of 250 mA. In the larger industrial cells, energy densities of up to 200 Wh/kg and specific capacities of 150 Ah/dm^3 may be obtained. On the other hand, a catalytic surface must be provided for efficient charge transfer at the oxygen cathode, and by its nature the electrode is susceptible to concentration polarization.

The history of metal–air batteries goes back over 100 years to the work of Maiche who modified a Leclanché cell by replacing the conventional MnO_2 cathode with a mixture of platinum and carbon powder. Successful commercial primary cells have ranged from the well known 500 Ah zinc–air wet cell developed for railway signalling in the 1930s to the more modern hearing aid batteries and high capacity/high current industrial primary systems. Secondary hybrid cells using air electrodes will be considered in Chapter 9.

The oxygen electrode

The oxygen electrode has been the subject of intensive study for many years. The electrode reaction is complex and is greatly affected by the electronic conductor and electrolyte used. In basic solution, it may be considered as a two stage process: only the first of these is reversible. The two steps may be written as

$$O_2(aq) + H_2O(1) + 2e \rightleftharpoons HO_2^-(aq) + OH^-(aq) \qquad (3.26)$$

and

$$HO_2^-(aq) + H_2O(1) + 2e \rightarrow 3OH^-(aq) \qquad (3.27)$$

In addition, other processes may occur such as the reaction of the hydroperoxide ion with the conductor to form metal–oxygen bonds which in turn may be reduced. The hydroperoxide ion may itself decompose to reform oxygen, etc. The potential of an oxygen electrode is invariably a mixed potential with a value of about 1.0 V on the standard hydrogen scale, at zero current drain.

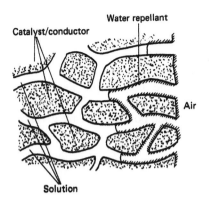

Fig. 3.29 Schematic view of the interphase at a porous matrix air electrode

Since the electrode reaction can only occur in the region where solid, liquid and gaseous phases come together, the construction of oxygen electrodes for practical cells is designed to maximize the interfaces between them. This may be achieved, for instance, by using porous nickel or carbon treated with metal or metal oxide catalysts (Kordesch electrodes). The pores are made accessible to both electrolyte solution and air (Fig. 3.29), and to prevent flooding and eventual leakage of the former, the surfaces of the electrode which are exposed to the air are impregnated with a water-repellent coating using paraffin wax or a synthetic polymer. It should be noted that oxygen electrodes of this type would be able to pass much larger currents if they were supplied with pure oxygen rather than air. Modern high performance air electrodes as used in button cells use laminated structures consisting of a Teflon water-repellent membrane with pore sizes optimized for particular discharge rates, bonded to a catalyst layer supported by a nickel mesh screen which acts as current collector, and finally a microporous polymer separator. Air is introduced to the electrode assembly either via access holes drilled through the cell case or by means of a diffusion membrane which ensures a uniform supply of air to the whole electrode surface. Regulation of air supply is important since cells can gain or lose water from or to the air when the relative humidity of the atmosphere is different from the equilibrium water vapour pressure of the electrolyte. Long service life requires limited air access, which in turn results in low rate capability. Oxygen consumption is 0.058 cm^3/C which corresponds to 5.8×10^{-5} cm^3/s for a 1 mA discharge. Prior to use, air access to the cell is blocked by an adhesive sealing tape.

The most common catalyst is MnO_2 or an oxide of manganese heat-treated in such a way as to optimize its catalytic properties and storage capability. Some cathodes rely solely on carbon.

The electrolyte for zinc-based cells is always caustic alkali. Calcium hydroxide is sometimes added to remove zinc ions as insoluble $CaZn_2O_3.5H_2O$. A caustic alkali electrolyte is effectively buffered against OH^-ion production by the oxygen cathode, so that OH^- concentration polarization is not serious. On the other hand, such an electrolyte can readily become contaminated with carbonate by reaction with the carbon dioxide of the air. NaOH is substituted by KOH in the zinc–air cells intended for low temperature use. Oxygen electrodes used with the near neutral electrolytes suitable for aluminium and magnesium anodes are more subject to polarization and are limited to lower current drain applications. Further details concerning oxygen electrode polarization are given in Chapter 9.

Anodes

Four metals have been studied extensively for use in this type of system, namely zinc, aluminium, magnesium and lithium. However, the last three metals suffer from severe corrosion problems during storage, and magnesium–air and aluminium–air cells are generally operated either as reserve systems in which the electrolyte solution is added to the cell only when it is decided to commence the discharge, or as mechanically rechargeable batteries which have replacement anode units available. The lower energy and power density of zinc are compensated for by the ease with which serious corrosion may be inhibited, so that zinc is by far the most commercially important anode in primary metal–air cells.

In early wet cells, lightly amalgamated solid zinc plates were used. Fig. 3.30 shows a cross-section of such a cell: the tapered shape of the zinc plates allowed for the higher current densities near the top of the air electrode. For button cells, the zinc anode is generally formed by mixing zinc powder with an electrolyte and gelling agent and compressing into a pellet, as in the zinc–silver oxide cell.

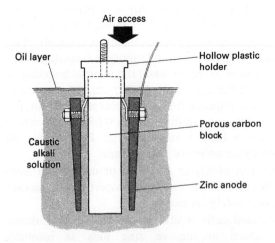

Fig. 3.30 Cross-section of the electrode assembly of an early 500 Ah mechanically rechargeable zinc–air wet cell used for railway signalling applications

In basic solution, unsaturated with zincate ions, the anode reaction may be written as

$$Zn(s) + 4OH^-(aq) - 2e \rightarrow ZnO_2^{2-}(aq) + 2H_2O(1) \qquad (3.28)$$

When the solution becomes saturated with zincate, zinc oxide is formed:

$$Zn(s) + 2OH^-(aq) - 2e \rightarrow ZnO(s) + H_2O(1) \qquad (3.29)$$

A number of techniques have been used to prevent degradation of battery performance caused by zinc oxide passivation. In the early wet cells, despite the penalty of reduced energy density, sufficient electrolyte was added to allow most of the zinc to dissolve. In more modern construction, anodes are made by compacting powdered zinc onto brass current collectors or by electrolytic reduction of pasted sheets, to form a porous mass with a high area/volume ratio. In this configuration the oxide does not significantly block further oxidation of the zinc. In addition, most of the electrolyte required may be incorporated within the pores. A carefully positioned current collecting grid is a necessity for such a cell since, by its nature, it is always anode limited.

The pros and cons of aluminium and magnesium anodes were discussed in Section 3.5. The corrosion problem is even more serious in metal–air cells since the electrolyte may be saturated with oxygen.

Typical cells

Zinc-based industrial primary cells range in size from 90 Ah cylindrical cells used mainly in telecommunications, hazard warning lights, etc., to 2000 Ah cells designed for inshore navigation beacons, standby power and railway track and signalling circuits.

The cell may be written as

$$Zn(s)|NaOH(aq)|C(s),O_2(g)$$

and the nominal OCV is 1.4 V.

With the larger cells, continuous discharge drains of 1 A and intermittent discharges of 2.5 A are possible. The internal resistance of such cells is low ($\approx 0.1 \ \Omega$). A typical discharge curve for a high current/high capacity cell is shown in Fig. 3.31.

Cells are also formed into batteries with nominal OCV of 7.2 V for electric fence activators. Alternatively, banks of 11 individual cells are used to operate 10 V railway signal motors: such applications require a current drain of 3 A for 5–10 s, perhaps 100 times per day. Multicell batteries are used in the US Space Shuttle programme for crew communication. Practical energy densities of up to 310 Wh/kg may be obtained.

A number of commercial zinc–air cells are available in reserve form; usually they require simply the addition of clean fresh or sea water to activate them.

In Fig. 3.32 a miniature 'air-depolarized' zinc cell is shown. Such cells

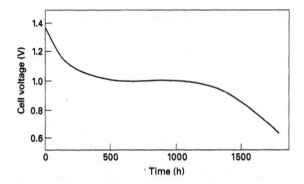

Fig. 3.31 Continuous discharge curve through 2 Ω load of a 750 Ah zinc–air cell. Such cells have a 5A intermittent pulse capability and good capacity retention. (By courtesy of SAFT (UK) Ltd.)

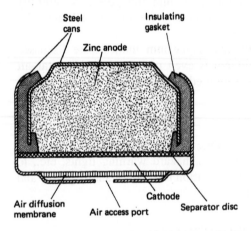

Fig. 3.32 Cross-section of zinc–air button cell. (By courtesy of Gould.)

were developed for use in hearing aids and similar devices. They have a similar configuration to the more conventional mercury or silver button cell, but a much higher capacity for the same volume, e.g. 560 Ah/dm^3 (190 Ah/kg). They are now also widely used for electronic pagers and for bone growth stimulation.

Zinc–air cells with electrolytes based on NH_4Cl and having a similar OCV are also manufactured:

$$Zn(s)|NH_4Cl(aq)|C(s),O_2(g)$$

The cell reaction can be written as

$$2Zn(s) + 4NH_4^+(aq) + 4Cl^-(aq) + O_2(g) \rightarrow$$
$$2Zn(NH_3)_2Cl_2(s) + 2H_2O(l) \tag{3.30}$$

Magnesium–air batteries have been developed as 'mechanically rechargeable' reserve batteries in modular form. In the General Electric multicell system, the air electrodes are integral with the battery casing, but the magnesium alloy anodes slot into guides and may be readily replaced. The electrolyte is sea water or a 7% NaCl solution. The so-called 'magnesium-inert cathode' cell is in practice a low efficiency magnesium-air cell. In this system the cathode is nickel-plated iron and the anode/cathode assembly is immersed in sea water to provide low currents in long-term low maintenance applications.

Aluminium–air cells are attractive in principle because of the high thermodynamic electrode potential and theoretical capacity of aluminium. Because of corrosion, however, the only commercial power sources based on this couple are either reserve systems or are mechanically rechargeable, i.e. the anodes are replaced after discharge. A typical cell may be written as

$$Al(s)|NaCl(aq)|M(s),O_2(g)$$

The anode reaction is written as

$$Al(s) \rightarrow Al^{3+}(aq) + 3e \qquad (3.31)$$

and the cathode reaction as

$$\tfrac{1}{2}O_2 + H_2O(1) + 2e \rightarrow 2OH^-(aq) \qquad (3.32)$$

so that the overall cell reaction is

$$4Al(s) + 3O_2(g) + 6H_2O(1) \rightarrow 4Al(OH)_3 \qquad (3.33)$$

The neutral electrolyte, often sea water, can be replaced by aqueous NaOH providing that the anode is alloyed with magnesium and tin. Alkaline aluminium–air cells have significantly greater power densities.

Aluminium–air cells were first developed for portable applications such as mooring lights, and for recharging nickel–cadmium and lead–acid storage batteries. They have been fabricated in many unusual designs, e.g. the concentric 'rope battery' which has an aluminium core surrounded by a separator and then the oxygen cathode. The 'rope' may be several hundred metres long and can provide 0.03 W/m for a period of 6 months on immersion in the sea.

More recently, aluminium–air systems have been developed for reserve power units, underwater propulsion and electric vehicles.

3.8 Magnesium reserve batteries

In previous sections of this chapter, magnesium anodes have been considered as replacements for zinc in Leclanché and air-depolarized cells. In sea water-activated reserve batteries, magnesium anodes are coupled with either silver chloride, lead chloride, manganese dioxide or, occasionally,

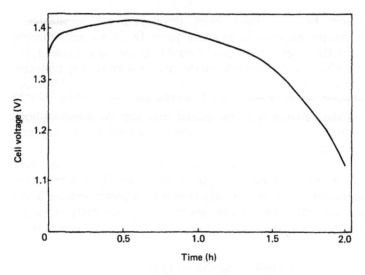

Fig. 3.33 Discharge curve for sea water-activated magnesium–silver chloride reserve cell

cuprous chloride, iodide or thiocyanate. In the case of silver chloride, the cell may be represented as

$$Mg(s)|sea\ water|AgCl(s),C(s)$$

and the cell reaction as

$$Mg(s) + 2AgCl(s) \rightarrow Mg^{2+}(aq) + 2Cl^-(aq) + Ag(s) \qquad (3.34)$$

OCV values fall in the range 1.6–1.9 V, and an energy density as high as 165 Wh/kg can be attained. Magnesium–lead chloride cells have energy densities up to 100 Wh/kg.

Sea water batteries can be activated simply by immersion or by more sophisticated automatic methods using forced flow. They are designed to provide currents for periods ranging from a few seconds to several hours (occasionally 1 or 2 days), most commonly with current drains from 150 mA to 25 A and voltages from 1.5 to 50 V. As the cells are inactive until water is introduced, they can be stored almost indefinitely in dry conditions at any temperature without problems connected with accidental short-circuiting or charging and without reduction in performance. They are designed to operate effectively over a wide range of salinity levels, temperature and depth. In Fig. 3.33, the discharge curve of a Mg–AgCl reserve cell is shown. The initial voltage rise is due to a combination of voltage delay at the anode, followed by the effects of a decrease in the electrolyte resistance as soluble magnesium chloride is formed. A schematic diagram of a sea water-activated battery is shown in Fig. 3.34.

Small cells are used to illuminate air-sea rescue beacons on life-jackets and life-rafts. Larger battery units with working voltages ranging up to several hundred volts are used for a variety of maritime applica-

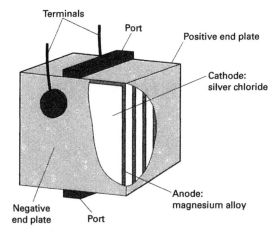

Fig. 3.34 Schematic diagram of a sea water-activated battery. (By courtesy of Ultralife.)

tions, including provision of motive power for torpedo motors, underwater countermeasures and sonobuoys. Batteries are also designed with stored electrolyte for airborne and air-deployed devices and land-based applications.

4 Primary lithium cells

Colin A Vincent

4.1 Introduction

The rapid development of high value electronics-based consumer products such as electric watches, calculators, 'smart' cards, remote controls, video games, etc. has made increasing demands on battery technology for systems with higher specific capacity, energy and power. Of all possible anode materials, lithium is perhaps the most attractive since it combines a favourable thermodynamic electrode potential with a very high specific capacity (3.86 Ah/g; 7.23 Ah/cm^3). As a result of its electropositive nature, lithium rapidly reduces water, and cells with lithium anodes generally employ non-aqueous electrolytes. In this chapter, systems operating at ambient temperatures and based on salts dissolved in aprotic organic solvents or in liquid cathodes will be discussed; lithium cells with molten salt, polymer or solid electrolytes will be reviewed in subsequent chapters, as will lithium secondary cells.

Research into lithium batteries began in the late 1950s. It was soon established that lithium metal was stable from a practical point of view in a number of electrolyte systems due to the formation of a passivating layer which prevented chemical corrosion but permitted cell discharge. This has been termed the *solid state interface* or *polymer electrolyte interface* and is considered in more detail below. The first commercial primary cell was introduced by SAFT in 1973. The initial emphasis was on military applications until various engineering and safety problems had been resolved. Within a further 10 years, primary cells with capacities ranging from 5 mAh to many thousands of Ah were available. In Fig. 4.1, the specific energy and power ranges of lithium-based ambient batteries are compared with those of other primary and secondary systems. The superior values of specific energy are evident. Specific power is, however, limited, mainly because of the relatively poor conductivity of the electrolytes. On

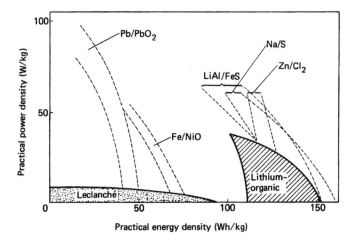

Fig. 4.1 Power density–energy density curves for practical battery systems

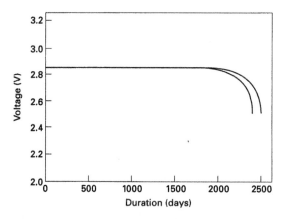

Fig. 4.2 Discharge of two 23 mm diameter × 2.5 mm thickness coin cells based on the Li–$(CF_x)_n$ couple, at 3 μA and room temperature. (By permission of National/Panasonic.)

the other hand, cells with lithium anodes have a number of other distinctive attributes that provide electrical characteristics which make them particularly suitable for supplying energy for electronics-based applications. The most important of these are:

- *High cell voltage.* Lithium cells commonly have OCV and working voltages of 3 V and may have values of 4 V. In addition to contributing to the high energy density of the cells, the number of cells in a battery pack can be reduced by a factor of 2 or 3 in comparison with aqueous primary cells.

- *Flat discharge.* It is straightforward to combine lithium with a cathode where the activity of the oxidized and reduced forms are invariant during discharge of the cell. Fig. 4.2 shows the discharge performance of Li/$(CF_x)_n$ cells over a 7 year period.

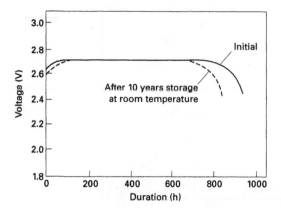

Fig. 4.3 Discharge performance of the cell in Fig. 4.2 at 0.2 mA before and after storage for 10 years at room temperature. (By permission of National/Panasonic.)

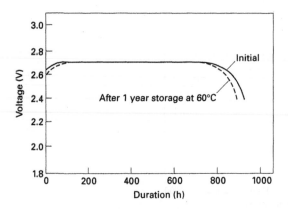

Fig. 4.4 Discharge performance of the cell in Fig. 4.2 at 0.2 mA before and after storage for 1 year at 60°C. (By permission of National/Panasonic.)

- *Long shelf life.* Self-discharge is minimized due to the formation of the passivating layer on lithium. Practical storage limits of 5–10 years at room temperature are readily achieved. In Figs 4.3 and 4.4, discharge curves for the same type of cell show that after 10 years of storage at 20°C the capacity has fallen by only 10%, while after 1 year at 60°C the capacity has not decreased significantly.

- *Wide operating temperature range.* Because of the low freezing point of suitable non-aqueous solvents, lithium cells can perform reasonably well down to temperatures of around −40°C. Their construction also allows them to be operated at 60°C and over.

Lithium primary cells have typical energy densities of up to 250 Wh/kg (500 Wh/dm^3), which are twice as high as the best conventional aqueous systems. Only the zinc–air cell can match or exceed these values, but it has other limitations as were discussed in Chapter 3.

Lithium systems may be classified according to the physical state of the positive electroactive material:

- solid cathode reagents: compounds with a negligibly small solubility in the electrolyte, e.g. $(CF_x)_n$, CuO, MnO_2, FeS;

- soluble cathode reagents: the only important example is sulphur dioxide, SO_2;

- liquid cathode reagents: the active species is in liquid form at the cell operating temperature, e.g. thionyl chloride, $SOCl_2$ and sulphuryl chloride, SO_2Cl_2.

Electrolytes are generally lithium salts dissolved in aprotic solvents or added to the liquid electroactive molecules as 'supporting electrolytes' to carry the current through the cell. Before describing a number of primary cells of technical and commercial importance, the general properties of the electrolytes, the lithium anode and the more common cathode materials will be discussed. Secondary cells are discussed in Chapter 7.

4.2   Electrolytes

Selection of the most suitable solute–solvent combination for a battery electrolyte involves consideration of the conductance of the resulting solution, its chemical and electrochemical stability and its compatibility with the electrode materials. In the case of lithium batteries, lithium inorganic salts dissolved in aprotic organic solvents have proved to be the most suitable electrolytes. The most commonly used solvents include cyclic esters (ethylene carbonate, propylene carbonate, γ-butyrolactone), linear esters, cyclic ethers (2-methyltetrahydrofuran, 1,3-dioxolane), linear ethers (1,2-dimethoxyethane), amides and sulphoxides: physical properties of these materials are listed in Table 4.1. A mixed solvent is sometimes preferred, since the properties of the electrolyte solution (conductance, viscosity, etc.) and its reactivity towards lithium can often be 'tailored' to give optimum performance. Higher conductivity is sometimes achieved by blending a low and a high polarity solvent, which generally show substantial negative viscosity deviations. Solvent mixtures are also used to modify the lithium surface layer, to remove impurities and inhibit side reactions. This is particularly important in the case of secondary systems (see Chapter 7) whose success depends largely on the suitability of the electrolyte for recharging the lithium electrode, and for subsequent charge retention. Because of the high voltage required during charge, solvent systems with very wide electrochemical stability windows are required. Choice of solute is limited mainly by solubility, which must be high so that the resulting solution will have sufficient conductivity and hence the cell

Table 4.1 Properties of some solvents commonly used in lithium–organic cells (at 25°C unless otherwise stated)

Solvent	Abbreviation	Molecular mass	m.p. (°C)	b.p. (°C)	Relative permittivity	Viscosity (cP*)	Density (g/cm³)
Acetonitrile	AN	41.05	−45.7	81.6	37.5	0.345	0.79
γ-butyrolactone	BL	86.09	−42	206	39.1	1.750	1.13
1,2-dimethoxyethane	DME	90.12	−58	85	7.05	0.455	0.86
N,N-dimethylformamide	DMF	73.10	−61.0	149	36.7	0.796	0.95
Dimethylsulphoxide	DMSO	78.13	18.5	189.0	46.7	1.960	1.10
Diethyl carbonate	DEC	118.13	−43	127	2.09	0.748	0.975
Dimethyl carbonate	DMC	90.08	3	90	3.09	0.585	1.07
1,3 dioxolane	DIOX	74.08	−97.2	105	7.6	0.677	1.06
Ethylene carbonate	EC	88.06	36.4	248	89.6 (40°C)	1.850 (40°C)	1.32 (40°C)
Methyl formate	MF	60.05	−99.0	31.5	8.5 (20°C)	0.340	0.97
2-methyltetrahydrofuran	MeTHF	86.12	−137.0	80	6.2	0.461	0.880
Nitromethane	NM	61.04	−28.5	101.0	35.9	0.620	1.14
Propylene carbonate	PC	102.09	−48.8	242	64.9	2.530	1.21
Tetrahydrofuran	THF	72.12	−108.0	65.0	7.4	0.457	0.848

* 1 cP = 0.001 kg m^{-1}s^{-1}

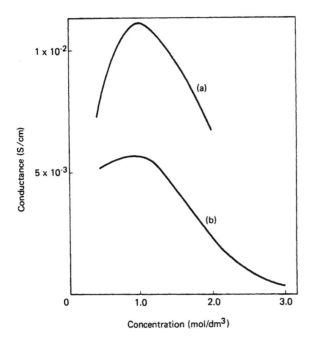

Fig. 4.5 Conductance of LiAsF$_6$ in γ-butyrolactone (a) and propylene carbonate (b) as a function of concentration. (By permission of the Electrochemical Society.)

will have a relatively low internal resistance. The most widely used salts are LiBr, LiPF$_6$, LiAsF$_6$, LiBF$_4$, LiClO$_4$ and LiAlCl$_4$, i.e. either simple salts or combinations of a lithium halide with a Lewis acid. More recently, a series of salts based on the sulphonate group, $-$ SO$_3^-$ have been developed. The best known, LiCF$_3$SO$_3$, lithium trifluoromethanesulphonate, sometimes referred to as 'lithium triflate' was commercialized by the 3M Company in the early 1980s and has been successfully used in primary (but not secondary) cells. More conductive solutions, due to lower ion pairing, may be obtained using LiN(SO$_2$CF$_3$)$_2$, known as LiTFSI or 'lithium imide', or LiC(SO$_2$CF$_3$)$_3$ which is called 'lithium methide'. Both of these salts are highly stable both from a thermal and from an electrochemical point of view.

The conductance of lithium salt solutions in aprotic solvents generally shows a maximum as the concentration of electrolyte is increased, as illustrated in Fig. 4.5. Such maxima can be interpreted on the basis of the opposing influence of an increasing number of charge carriers on the one hand, and increasing viscosity and increasing ion association with the formation of non-conducting ion pairs, on the other. It has been shown that the conductance can be increased by the addition of crown ethers, such as 12–crown–4:

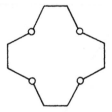

or linear polydentate ethers such as 1,2 dimethoxyethane. This can be attributed to a decrease in ion-association due to shielding of the charge on the lithium ions.

In Table 4.2, the solubility of $LiAsF_6$ in a number of solvents is given. In methyl formate (MF), $LiAsF_6$ dissolves to give one of the most conductive organic electrolyte solutions known: at 25°C a solution of 2 mol/dm^3 has a conductance of 4.5 S/m. While this combination can be considered as a suitable electrolyte for 'high rate' cells, it suffers from some decomposition with gassing in the presence of lithium unless stabilized with materials such as $(CH_3)_4NBF_4$. $LiAlCl_4$ is the solute which is most widely used to increase the conductance of inorganic liquid cathode materials such as $SOCl_2$ and SO_2Cl_2. Typical conductance values for these solutions are given in Table 4.3.

The solvents used in lithium batteries are generally thermodynamically unstable in the presence of lithium. The low lithium corrosion rate and consequent good shelf life actually experienced with sealed cells is due to the formation of a protective film on the surface of the metal. The practical stability of the electrolyte solutions in the presence of lithium depends on

Table 4.2 Maximum solubility of $LiAsF_6$ at 25°C

Solvent	Concentration (mol/dm³)
Acetonitrile	1.60
Dimethylformamide	4.68
Propylene carbonate	4.66
Water	3.28

Table 4.3 Conductivity of some $LiAlCl_4$ solutions in organic solvents at 25°C

Solvent	Concentration (mol/dm³)	Conductance (S/m)
$SOCl_2$	2.0	2.0
SO_2Cl_2	2.0	1.0

their purity and, in particular, on low water content. A number of procedures have been proposed for the purification and drying of solvents, including operations such as pre-electrolysis with platinum electrodes. In most cases it has proved sufficient to place the solvent in contact with molecular sieves for a few days before fractional distillation.

The electrochemical stability of a solution can be defined as the voltage range within which it does not undergo detectable electrolytic decomposition, for a particular pair of electrodes. In the case of a practical primary cell, such a range must be wider than the OCV. For secondary systems, the value of the voltage applied to the cell during recharge must also be considered. The nature and history of the electrodes have an important influence on the decomposition voltage. This illustrates the importance of kinetic over thermodynamic factors. It is often difficult to establish whether a decomposition current is due to reactions involving solvent or solute, but in practice a marked influence of the nature of the solute is often observed.

4.3   The lithium anode in primary cells

The properties of lithium are summarized in Table 4.4. The metal is both ductile and malleable. It is harder than the other alkali metals, but softer than lead. It shows a tendency to stick to various materials under moderate pressure. In humid air the freshly cut shiny surface of lithium rapidly turns greyish-white to grey or black, due to a reaction with oxygen and nitrogen. However, if the humidity is maintained at a low enough level (< 2% at 20°C), there is no apparent tarnishing for periods of hours. 'Dry rooms' can be constructed for the assembly of lithium batteries. The handling of lithium is comparatively safe in comparison with the other alkali metals; for example, it will not spontaneously ignite, unless in finely divided form and in contact with water.

Table 4.4 Physical properties of lithium

Atomic weight	6.94
Melting point	180.5°C
Boiling point	1347°C
Density	0.534 g/cm^3 at 20°C
Heat of fusion	3.001 kJ/mol
Heat of vaporization	147.1 kJ/mol
Resistivity	9.446 × 10^{-6} Ω cm at 20°C

Film formation on lithium

The reaction of lithium with the electrolyte to form a surface film significantly modifies its behaviour. On the one hand, the film confers chemical
stability and useful shelf life on the system. On the other, it is responsible
for greatly depressed exchange currents and the consequent phenomenon of
voltage delay, as discussed in Chapter 3 in connection with magnesium
aqueous batteries. It is convenient to discuss separately film formation with
insoluble and with liquid and soluble cathode systems.

Cells with insoluble cathodic reagents

In these cells, provided that the solubility of the cathode material is very
low, the solvent itself is principally responsible for film formation although
the anion of the salt is often also involved. Lithium was originally thought
to react with propylene carbonate (PC) to form gaseous propene and
lithium carbonate as follows:

$$CH_3 - CH - CH_2(l) + 2Li(s) \rightarrow CH_3 - CH = CH_2(g) + Li_2CO_3(s) \quad (4.1)$$

Propene evolution has been observed on lithium amalgams and also at
platinum surfaces connected to lithium electrodes. It is now known that the
reaction on a lithium surface is much more complex, and involves poly-
merization of the PC and the formation of lithium alkyl carbonates which
subsequently hydrolyse with trace water to form $LiCO_3$. The film structure
is also known to be complex. Ether-derived films have also been studied
and have been shown to be largely alkoxide-based with evidence for
various polymerized species. Passive films have also been observed with
solvents such as BL, AN, MF, THF, etc. The most important characteristics
of an optimized film are that it is adherent, insoluble, thin and has negligible electronic conductivity. Such a film will passivate the lithium and
prevent further corrosion during an extended shelf life. However, in order
to permit the cell to discharge when connected to a load, the film must have
a high lithium ion mobility, i.e. it must behave as a lithium ion-conducting
solid electrolyte.

On drawing current from a passivated lithium anode, polarization may be
at first severe, but the voltage recovers fairly rapidly (Fig. 4.6). Initially,
charge transfer at the anode is limited by lithium ion transport through a
thin or imperfect section of the interfacial film. This process progressively

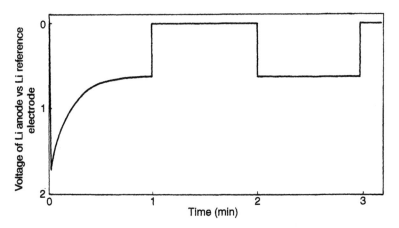

Fig. 4.6 Voltage recovery of a lithium anode at $-20°C$ in 1 mol/dm^3 LiClO$_4$ in PC versus a lithium reference electrode. Current density $= 10$ mA/cm^2. (By permission of the Electrochemical Society: N. Margalit and H.J. Canning, *Proceedings of the symposia on power sources for biomedical implantable applications and ambient temperature lithium batteries*, eds B.B. Owens and N. Margalit, 1980, p. 339.)

disrupts the film and partially removes it, thus allowing the discharge voltage to rise rapidly to its steady value. If the cell is now returned to open circuit, and a second current pulse is subsequently withdrawn, no voltage delay is observed if the period on open circuit is too brief for reformation of the film. In commercial lithium batteries of this type, recovery from voltage delay is almost instantaneous under normal working conditions, but is slower at low temperatures.

Cells with liquid or soluble cathodic reagents

The formation of passivating films on lithium in contact with liquid or soluble cathodic reagents is a prerequisite for the construction of a practical cell. The film acts in the same way as a separator, preventing further direct chemical reaction of lithium and the cathodic reagent. However, film formation involving the action of SO$_2$, SOCl$_2$, etc. on lithium is considerably more complex and may produce much more severe voltage delay characteristics than in the case of insoluble cathodes described above.

When SO$_2$ dissolved in AN is brought into contact with lithium, a layer of lithium dithionite is formed, following the same reaction scheme as the normal cell reaction:

$$2Li(s) + 2SO_2(diss.) \rightarrow Li_2S_2O_4(s) \tag{4.2}$$

If the solvent is a PC/AN mixture, the film constitution is complex and contains Li$_2$CO$_3$ in addition to Li$_2$S$_2$O$_4$. This latter film is probably more coherent or compact since the voltage delay is greater than for the pure dithionite film. With SOCl$_2$ and SO$_2$Cl$_2$ containing LiAlCl$_4$ a particularly severe voltage delay is observed, especially after prolonged storage, as shown in Fig. 4.7. The film consists primarily of LiCl crystals whose

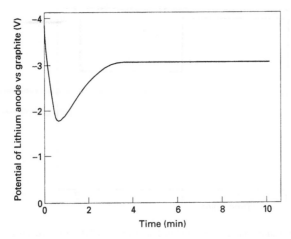

Fig. 4.7 Initial discharge behaviour of a Li(s)|LiAlCl$_4$,SO$_2$Cl$_2$(l)|C(s) cell after prolonged storage at room temperature. Current density at lithium electrode = 5 mA/cm^2

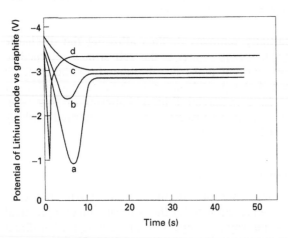

Fig. 4.8 Effect of electrolyte composition on discharge behaviour of Li(s)|(electrolyte, 1 mol/dm^3), SOCl$_2$(l)|C(s). (a) AlCl$_3$.LiCl; (b) AlCl$_3$.Li$_2$S; (c) AlCl$_3$.Li$_2$O; (d) AlCl$_3$.LiF. Current density at lithium electrode = 6.4 mA/cm^2. Storage: 15 days at 25°C or 7 days at 70°C. (By permission of Bressan, De Guibert and Feuillade, Extended Abstracts, ISE meeting, 1980, vol. 2, p. 737.)

growth and morphology follow a complex pattern and which are affected substantially by additives such as AlCl$_3$, S$_2$Cl$_2$, H$_2$O and by other electrolyte variables. A number of investigators have shown how film growth in these systems can be controlled and voltage delay reduced by the careful selection of electrolyte components. The advantage of neutralizing the acidic AlCl$_3$ with stronger Lewis bases, e.g. Li$_2$O rather than LiCl, has been demonstrated. The effect of the change of the Lewis acid–base properties of the electrolyte on the voltage delay are shown in Fig. 4.8. The influence of alloying the lithium anode has also been studied.

In practical terms, the twin objectives of protecting the lithium from corrosion while avoiding unacceptable levels of voltage delay can be considered to have been met. However, the detailed mechanisms of film formation and disruption are still matters of some controversy. In particular, the interaction of thin films formed rapidly on lithium surfaces exposed to the atmosphere with the thicker films formed by subsequent reaction with the cathodic reagent is not well understood.

4.4   Cathode materials and lithium primary cells

Lithium primary cells may be conveniently divided into three categories according to the type of cathodic reagent used, as suggested in Section 4.1. Cells are manufactured in a number of forms of which button or coin cells form one important group, while cylindrical cells of standard dimensions form the second. For the latter, bobbin-type design is used for high capacity and energy density, while a spiral wound ('jelly roll') configuration (Fig. 4.9) is used for higher rate applications. Specialized cells such as the miniature cylindrical pin-type cell (Fig. 4.10) or the prismatic defibrillator battery (Fig. 4.11) are also available. Film or 'paper' batteries are also manufactured, which have a total thickness of 0.5 mm or less. For larger batteries, flat parallel plates are used in prismatic containers. Military standby power cells have capacities up to 20 000 Ah.

Solid cathode systems

Four main groups of compounds may be distinguished: polycarbon fluorides, oxosalts, oxides and sulphides.

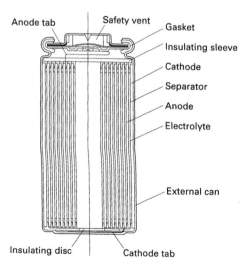

Fig. 4.9 Spiral wound ('jelly roll') configuration

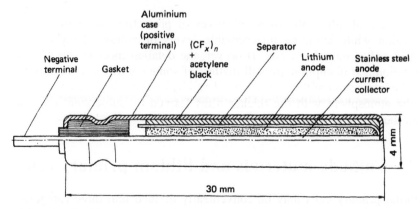

Fig. 4.10 Lithium–$(CF_x)_n$ cylindrical cell (based on the design of Matsushita (BR435), by permission.)

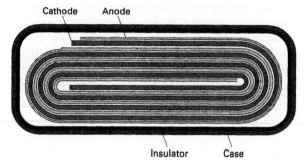

Note: Separator not shown

Fig. 4.11 Cross-sectional view (from the top) of a prismatic high power lithium–silver vanadium oxide battery used to power an implantable cardioverter defibrillator (ICD). (By permission of Medtronic.)

Polycarbon fluorides

The coupling of lithium with free halogens gives rise to cells of exceptionally high specific energy. Unfortunately the chemical reactivity of the components is generally too high. (The solid state lithium–iodine cell is, however, a commercial success, and prototype solid state lithium–bromine cells have been successfully tested (Chapter 9).) In the case of fluorine, the problem has been overcome by effectively immobilizing the fluorine in a graphite host. Polycarbon fluorides of general formula $(CF_x)_n$ can be obtained by direct fluorination of carbon black, or other carbon varieties at high temperatures. For fluorine compositions in the range $0.4 < x < 1.0$, such substances have very high specific energies (e.g. 2600 Wh/kg for $x = 1$). For $x < 1$, the materials are electronic conductors. Lithium cells with polycarbon fluoride cathodes have OCV values in the range of 2.8–3.3 V, depending on the exact formulation of the cathode material. Since $x \approx 1$ for commercial cathode material, cells are generally referred to as Li–$(CF)_n$. Carbon is added to such cathodes to increase the electronic conductivity.

A typical cell may be written as

$$Li(s)|LiBF_4.BL–DME|(CF_x)_n(s)$$

with associated cell reaction

$$nxLi(s) + (CF_x)_n(s) \rightarrow nC(s) + nxLiF(s) \qquad (4.3)$$

It is believed that the discharge mechanism involves the formation of an intermediate lithium 'intercalation' compound in which both lithium and fluorine are situated between the carbon layers of the graphitic structure. The carbon formed is graphitic and improves the cell performance as the discharge progresses, leading to a high cathode utilization – close to 100% for low currents. The lithium fluoride precipitates.

The $(CF_x)_n$ cathode is stable in contact with organic electrolyte systems, and the cells have a satisfactory shelf life. In common with other lithium power sources, polycarbon fluoride-based cells may suffer from voltage delay, but only under severe discharge conditions.

Cells based on polycarbon fluorides are manufactured commercially in a number of forms. The system was developed first by Matsushita Electrical Industrial Co. in Japan, and cells for military applications have been produced in the USA by Eagle Picher and Yardney Electric. The spiral wound cylindrical cells have the largest capacity and are used in portable radio transceivers, surveying equipment, computer memory back-up, etc. The so-called 'inside-out' cells have a hollow cylindrical cathode and a central lithium anode, all enclosed in a light aluminium case. A cross-section of such a cell based on Matsushita design was shown in Fig. 4.10: the long thin form ('pin-type') is an advantage in particular applications such as fishing floats. The same company also manufactures spiral wound cylindrical cells with capacities ranging from 600 to 5000 mAh and two-unit battery packs for automatic cameras. These are able to work over a very wide temperature range (−40 to +85°C) and can supply pulse currents of over 1 A to operate the flash. Discharge curves at 20 μA and pulse discharge voltages at 1 A are shown for this unit as a function of temperature in Figs 4.12 and 4.13, respectively. Finally, the popular button cell is produced mainly for use in electronic watches and pocket calculators. Discharge curves of these cells under various loads are shown in Fig. 4.14. Table 4.5 gives the specifications of typical cells of each type. The energy density of lithium–polycarbon fluoride cells depends on cell size, construction and discharge rate, but practical values of 250–300 Wh/kg and over 600 Wh/dm^3 give them among the highest energy densities of any commercial primary cells. The cells also have good storage characteristics, with a self-discharge rate of no more than 0.5% per annum under normal conditions. This property makes lithium–polycarbon fluoride cells very

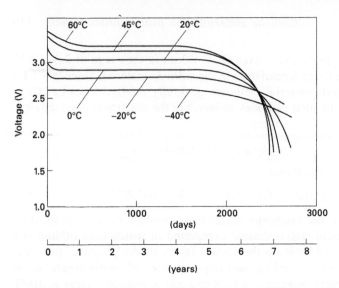

Fig. 4.12 Discharge characteristics of a spiral wound cylindrical cell (17 mm diameter × 33.5 mm height) based on the Li–$(CF_x)_n$ couple, at 20 µA over a wide range of temperatures. (By permission of National/Panasonic.)

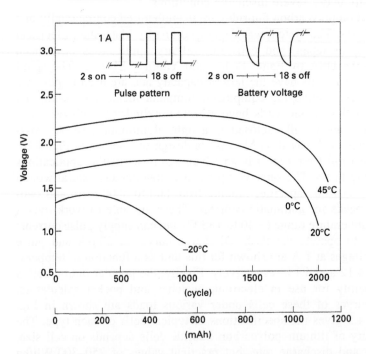

Fig. 4.13 Pulse discharge characteristics of the cell in Fig. 4.12 for 1A pulses (2 s 'on'; 18 s 'off') for temperatures from −20 to 45°C. (By permission of National/Panasonic.)

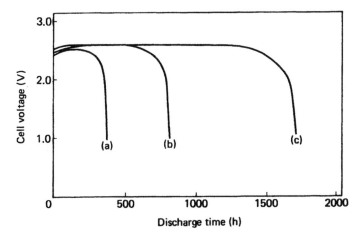

Fig. 4.14 Discharge curves of lithium–$(CF_x)_n$ button cells under various loads at ambient temperature: (a) 5 kΩ; (b) 13 kΩ; (c) 30 kΩ. (By permission of Elsevier Sequoia, SA.)

Table 4.5 Specifications of some commercial Li–$(CF_x)_n$ cells (National Matsushita Electrical Industrial Co., Japan)

	Button	Inside-out	Spiral
OCV	3.0	3.0	3.0
Nominal capacity (mAh)	150	40	5000
Energy density (Wh/kg)	140	140	320
Diameter (mm)	23	4.2	26
Height (mm)	2.5	35.9	50
Weight (g)	3.1	0.85	47.0

suitable for low-drain applications where a few microamps, say, must flow for 5–6 years (Fig. 4.2).

Oxosalts

Silver, copper and other oxosalts have been extensively studied as cathodes in laboratory cells; commercial power sources, principally for pacemakers, using silver chromate were manufactured until the 1980s, and silver vanadate or silver vanadium oxide ($Ag_2V_4O_{11}$), first reported by workers at Wilson Greatbatch Ltd, is currently used as cathode in implantable cardiac defibrillator batteries.

The main discharge process for lithium cells based on silver chromate is

$$2Li(s) + Ag_2CrO_4(s) \rightarrow 2Ag(s) + Li_2CrO_4(s) \qquad (4.4)$$

for which the cell

$$Li(s)|LiClO_4,PC|Ag_2CrO_4(s),C(s)$$

Table 4.6 Practical performance characteristics of some oxosalts in cells with LiAsF$_6$/BL electrolyte at 25°C, based on a two-electron reduction

Salt	OCV vs Li$^+$/Li (V)	Current density (mA/cm^2)	Specific capacity (Ah/g)	Specific energy (Wh/g)	(Wh/cm^3)	Utilization to 1V cut-off (%)
Ag$_2$WO$_4$	3.20	1.00	0.10	0.29	1.48	95
		8.00	0.07	0.20	1.02	75
Ag$_2$MoO$_4$	3.00	1.00	0.12	0.36	1.78	95
		8.00	0.06	0.13	0.64	50
CuWO$_4$	3.20	1.00	0.16	0.35	1.64	90
		5.00	0.16	0.28	1.35	60
		8.00	0.13	0.15	0.72	40
CuMoO$_4$	3.25	1.00	0.29	0.46	1.36	98
		5.00	0.21	0.23	0.68	90
		8.00	0.13	0.19	0.58	50

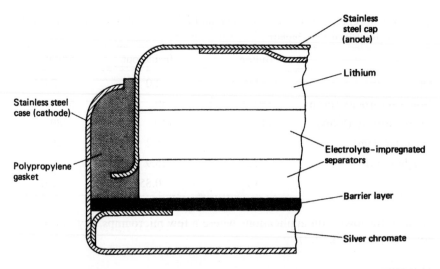

Fig. 4.15 Cross-section of a typical lithium–silver chromate button cell. (By courtesy of SAFT Gipelec.)

has a nominal OCV of 3.5 V. A second reduction process follows at about 2.5 V, associated with the reduction of Cr(VI). Other oxosalts behave in a broadly similar manner. In Table 4.6, the practical energetic characteristics of a number of cathode materials coupled with lithium in LiAsF$_6$–BL cells are compared at two current densities.

Silver chromate-based cells were manufactured in button and rectangular (prismatic) form in a number of sizes. The energy density of such complete systems is estimated as 200 Wh/kg or 575 Wh/dm^3, to a 2.5 V cut-off. A schematic cross-section of a typical button cell of this type is shown in Fig.

4.15. An exploded view of a rectangular cardiac pacemaker cell which was produced by SAFT is given in Fig. 4.16.

Silver vanadium oxide, $Ag_2V_4O_{11}$, is a semiconducting vanadium oxide bronze which adopts at least two related structures based on V_4O_{11} layers with silver atoms located between them (Fig. 4.17). The open structure allows facile diffusion of lithium ions. $Ag_2V_4O_{11}$ can be lithiated with up to seven lithium atoms to form $Li_7Ag_2V_4O_{11}$.

Two regions of flat OCV occur at $0 < x < 2$ and $3 < x < 5.2$ for the Li–$Li_xAg_2V_4O_{11}$ cell with sloping regions at compositions in the ranges $2 < x < 3$ and $5.2 < x < 7$. There is still some uncertainty concerning the electrode reaction, but it is likely that the first step involves reduction of silver from I to 0, and that subsequent steps are based on the reduction of vanadium V to IV and from IV to III. The variation of OCV with depth of discharge is a useful feature for indicating the remaining charge in the cell.

Cells for implantable cardiac defibrillators require a capacity of 1 Ah and the capability of supplying current pulses of 4–5 W for periods of

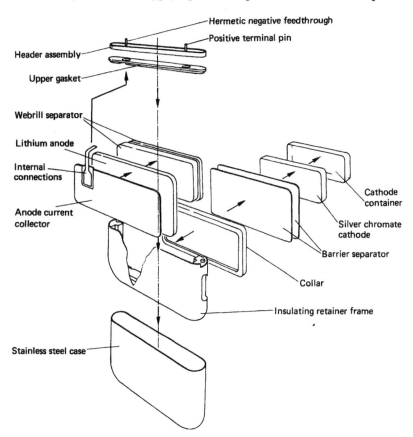

Fig. 4.16 Exploded view of rectangular (prismatic) lithium–silver chromate pacemaker cell. (By courtesy of SAFT Gipelec.)

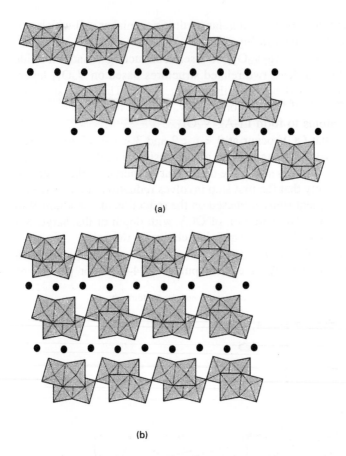

(a)

(b)

Fig. 4.17 The two structures of $Ag_2V_4O_{11}$ in the a–c plane: (a) type I, (b) type II. The distorted octahedra represent the V_4O_{11} layers, while the dark circles represent the silver atoms. The structures are related by translation of the V_4O_{11} layers between one of two stacking arrangements. (By permission of Medtronic.)

10–15 s. Clearly, reliability, safety and predictability of long-term performance are at a premium for such an application.

A number of cell designs have been employed. Figure 4.11 showed a prismatic design in which the electrodes encased in a Celgad 4560 separator are wound in a flattened coil. Standard electrolytes are used, such as $LiAsF_6$ or $LiClO_4$, in PC/DME. Titanium and niobium are used for current collector and the feedthrough pin of the glass-to-metal seal, respectively, because of their high corrosion resistance. The cell is cathode limited and is hermetically sealed in a stainless steel case which is connected to the lithium electrode.

Oxides

The discharge reaction for lithium–metal oxide cells is almost always more complex than the formal displacement process:

Table 4.7 Theoretical capacities for a number of simple displacement reactions involving oxides

Active cathode material	Product	Electrons involved	Capacity	
			Ah/g	Ah/cm^3
Bi_2O_3	Bi	6	0.350	3.070
CuO	Cu	2	0.670	4.260
MnO_2	Mn_2O_3	1	0.310	1.550
Pb_3O_4	Pb	8	0.310	2.850

$$2Li(s) + MO(s) \rightarrow Li_2O(s) + M(s) \tag{4.5}$$

For example, Bi_2O_3 and PbO give rise to intermetallic compounds such as Li_3Bi and LiPb, while MoO_3 and V_2O_5 form ternary phases. In many cases (e.g. for the MnO_2 cathode), the discharge mechanisms are still not fully understood. In Table 4.7, theoretical capacities for a number of oxides are compared, assuming a straightforward displacement reaction. While a wide variety of metal oxides have been studied, cathodes of commercial primary cells are mainly confined to the oxides of manganese and copper. Nb_2O_5, V_2O_5, V_6O_{13} and CoO_2 are used as positive plates in rechargeable lithium batteries; these will be considered in Chapter 7.

Two categories of cell may be distinguished: 'high voltage' and 'voltage compatible' cells. The latter term refers to the fact that lithium cells with discharge voltages of about 1.5 V can readily replace the more conventional miniature aqueous cells for which much electrical equipment has been designed.

The lithium–manganese dioxide high voltage cell introduced by Sanyo in 1975 is now manufactured by numerous companies, using specially prepared high density MnO_2. The OCV is in the range 3.0–3.5 V and typical

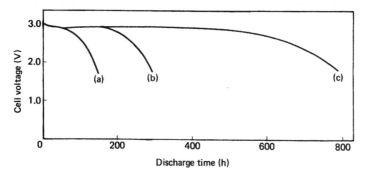

Fig. 4.18 Discharge curves of lithium–manganese dioxide button cells (Varta CR 2025) under various loads at ambient temperature: (a) 2.7 kΩ; (b) 5.6 kΩ; (c) 15 kΩ. (By courtesy of Varta.)

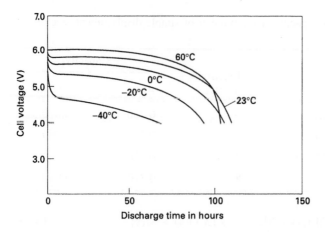

Fig. 4.19 Discharge characteristics of two series-connected–spiral-wound cylindrical cells (15.2 mm diameter × 40 mm height) based on the Li–MnO$_2$ couple, at 14 mA over a wide range of temperatures. (By permission of Sanyo.)

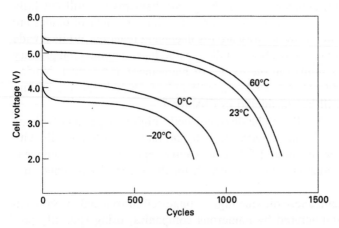

Fig. 4.20 Pulse discharge characteristics of the cell combination in Fig. 4.19 for 1.2 A pulses (3 s 'on'; 7 s 'off') for temperatures from −20 to 60°C. (By permission of Sanyo.)

discharge curves are shown in Fig. 4.18 for button cells and in Figs 4.19–4.21 for continuous and pulse loads on high rate Sanyo 2 CR 5 pack.

Low rate flat or 'coin' cells are manufactured with rated capacities from 32 to 1000 mAh with energy densities of 100–250 Wh/kg (400–750 Wh/dm^3). They are used for watches, calculators and cameras, and are capable of supplying occasional short high current pulses, e.g. to illuminate a miniature tungsten bulb in order to enable the time to be read at night on a travelling clock with an LCD display. Fig. 4.22 shows the behaviour of a cell under such a load.

Bobbin-type and spirally wound cylindrical cells are now fabricated by a number of manufacturers. The former are used for low-drain long-term applications such as computer memory retention and have a useful life of at

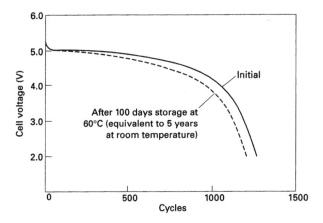

Fig. 4.21 Pulse discharge characteristics of the cell combination in Fig. 4.19 for 1.2 A pulses (3 s 'on', 7 s 'off') at 23°C before and after storage for 100 days at 60°C. (By permission of Sanyo.)

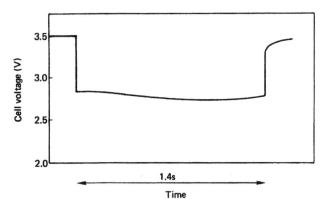

Fig. 4.22 Response to high current discharge pulse of a typical lithium–manganese dioxide watch battery (LM 2020). Current = 6 mA. (By courtesy of SAFT Gipelec.)

least 10 years. The latter are generally used for applications requiring currents in the range 10–100 mA, but a typical commercial cell (Sanyo CR 15400) weighing 17 g and having a volume of 30 cm^3 can sustain a continuous discharge current of 1.5 A and a pulse discharge of 3.5 A. Such batteries are often used in applications which involve electric motors in automatic cameras, toys, etc. The high rate cells require protection against short circuit and cell reversal; this is often provided by incorporating a series positive temperature coefficient resistor. Various venting mechanisms are also provided. Matsushita have designed a metallic seal plate which deforms when the cell pressure rises until it is pierced by a sharp steel strip, releasing the pressure through gas vent holes in the upper cap. The use of 'shutdown' separators has also been developed over recent years. Microporous polyolefin separators melting at 125°C when the cell is short-circuited rapidly reduce the current.

Prismatic 9 V three-cell batteries, and 6 V two-cell 'ultrathin' batteries are also manufactured. The former is designed as a high capacity replacement for aqueous zinc primaries and the latter for 'smart' cards.

The lithium–copper oxide cell is 'voltage compatible' (OCV ≈ 1.5 V), i.e. it may be used as a direct replacement for conventional Leclanché or alkaline zinc cells. CuO has a particularly high volumetric capacity (4.2 Ah/cm^3) so that cells are characterized by high specific energy – 300 Wh/kg (700 Wh/dm^3). The discharge curve shows a single step which may be attributed to the simple displacement reaction:

$$2Li(s) + CuO(s) \rightarrow Li_2O(s) + Cu(s) \qquad (4.6)$$

For some high temperature, low current discharge conditions a double plateau has been reported which may indicate a two-stage reduction of Cu II as in the case of CuS considered below.

It has been shown that cathode utilization and shelf life are improved by incorporation of about 1 m/o of Li_2CO_3 or LiOH in the CuO, followed by heat treatment. ('m/o' is used to represent the molar percentage of a component in a mixture.) The electrolyte used varies from manufacturer to manufacturer: $LiClO_4$ in 1,3 dioxolane has proved to be a very satisfactory system. Once again, self-discharge is very low. In Fig. 4.23, discharge curves after accelerated shelf testing at 70°C are shown. The cells behave well over a wide temperature range at moderate current drain. In Fig. 4.24, the capacity of this type of cell is compared with an aqueous alkaline zinc cell of similar dimensions, under various conditions. The system has also been satisfactorily tested in resin-encapsulated cells at temperatures as high as 150°C. The cells are manufactured in cylindrical form with theoretical capacities in the range 500–3900 mAh. A comparison of practical discharge capacities of lithium–copper oxide button cells and a number of aqueous alternatives is given in Fig. 4.25.

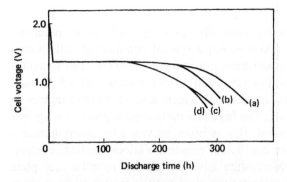

Fig. 4.23 Discharge curves of lithium–copper oxide button cells (LC 01) after accelerated shelf testing at 70°C: (a) fresh cell; (b) after 6 months at 70°C; (c) after 12 months at 70°C; (d) after 18 months at 70°C. (By courtesy of SAFT Gipelec.)

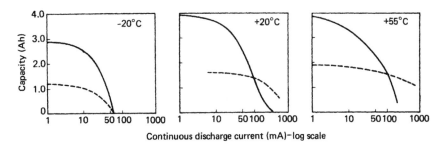

Continuous discharge current (mA)-log scale

Fig. 4.24 Capacity of lithium–copper oxide button cells (LC 01) as a function of discharge current (solid lines). Dashed lines indicate the characteristics of alkaline manganese cells of similar dimensions. (By courtesy of SAFT Gipelec.)

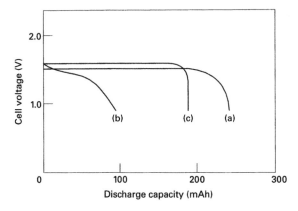

Fig. 4.25 Comparison of discharge curves at ambient temperature of voltage-compatible lithium–copper oxide button cells and conventional aqueous cells: (a) lithium–copper oxide; (b) alkaline manganese; (c) zinc–silver oxide. Load = 75 kΩ

The lithium–copper oxyphosphate cell has similar features to the lithium–copper oxide cell, but has a somewhat higher voltage. The cell reaction is written as

$$Cu_4O(PO_4)_2(s) + 8Li(s) \rightarrow 4Cu(s) + Li_2O(s) + 2Li_3PO_4(s) \quad (4.7)$$

Both of these copper-based lithium primaries are manufactured as button and bobbin-configured cylindrical cells. Copper oxyphosphate cells find particular application in high temperature environments.

Other oxide 'voltage compatible' cathodes under study include a mixed bismuth and lead oxide of composition $Bi_2Pb_2O_5$, and a similar mixed bismuth and copper oxide, Bi_2CuO_4.

Sulphides

Metal sulphides have the advantage over the corresponding oxides that most of them are good electronic conductors and hence sulphide-based cathodes do not usually require the addition of carbon.

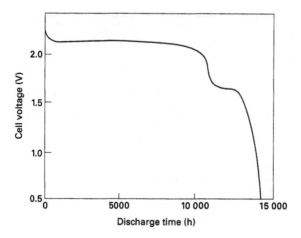

Fig. 4.26 Discharge curve of a lithium–cupric sulphide pacemaker cell at 37°C under a load of 12.3 kΩ. (By permission of the Electrochemical Society: A.J. Cuesta and D.D. Bump, *Proceedings of the symposia on power sources for biomedical implantable applications and ambient temperature lithium batteries*, eds B.B. Owens and N. Margalit, 1980, p. 95.)

Batteries based on cupric sulphide cells (three in series) were originally developed and used with cardiac pacemakers. Reduction of CuS takes place in two stages:

$$2CuS(s) + 2Li(s) \rightarrow Cu_2S(s) + Li_2S(s) \qquad (4.8)$$

and

$$Cu_2S(s) + 2Li(s) \rightarrow 2Cu(s) + Li_2S(s) \qquad (4.9)$$

so that the discharge curve has two stages with plateaus at 2.12 and 1.75 V. The Cordis pacemaker cells were anode limited so that about 80% of the cell capacity has been delivered when the fall in cell voltage occurred (Fig. 4.26). As in the case of the similar lithium–silver chromate cell described above, this voltage drop was used to indicate when battery replacement was required. The electrolyte in this cell was $LiClO_4$ dissolved in a mixed solvent, of which the major constituent was 1,3 dioxolane.

Cells based on cuprous sulphide were also developed: the continuous discharge characteristics of a Ray-O-Vac thin coin cell are shown in Fig. 4.27.

The commercially most important voltage compatible lithium primary cell uses iron disulphide, FeS_2, as cathode. The overall cell reaction may be written as

$$4Li(s) + FeS_2(s) \rightarrow Fe(s) + 2Li_2S \qquad (4.10)$$

but intermediate species such as FeS may be formed during the discharge.

Li–FeS_2 cells were manufactured as button cells, and are now produced as spiral wound cylindrical cells (Fig. 4.28). The latter are currently being heavily marketed worldwide by the Eveready Battery Co, Inc. (especially

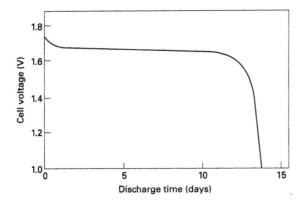

Fig. 4.27 Discharge curve of a lithium–cuprous sulphide thin coin cell under a load of 12.5 kΩ. (By courtesy of INCO.)

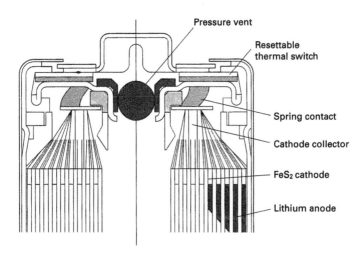

Fig. 4.28 Schematic cross-section of the upper part of the 'Energiser® L 91' spiral wound Li–FeS$_2$ cylindrical cell, showing the positive temperature coefficient (PTC) resettable safety switch, and the pressure relief vent. (By permission of Eveready.)

the AA size), as high capacity replacements for alkaline zinc equivalents. Such cells are equipped with positive temperature coefficient thermal switches, as discussed above for Li–MnO$_2$ cells, and pressure release devices.

Lithium–iron disulphide cells have an OCV of about 1.8 V, but are generally discharged at 1.2–1.5 V under heavy drain conditions (Fig. 4.29). They provide superior discharge characteristics in comparison with alkaline manganese cells for high rate discharge (Fig. 4.30). They are also capable of service over a much broader temperature range, typically −40 to +60°C. They have a wide application in high drain consumer

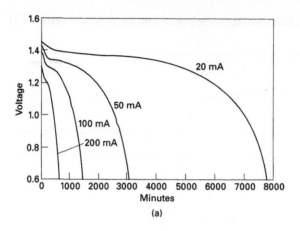

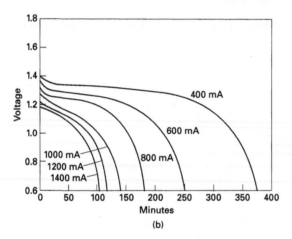

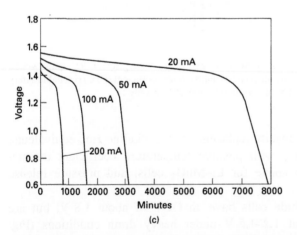

Fig. 4.29 Discharge characteristics of the 'Energiser® L 91' cell in Fig. 4.28. (a) and (b) are at 21°C and (c) is at −20°C. (By permission of Eveready.)

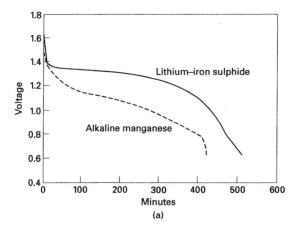

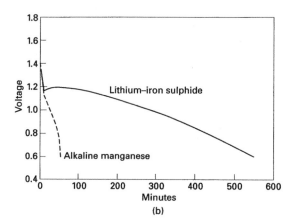

Fig. 4.30 Comparison of the discharge curves of AA-size lithium–iron sulphide and alkaline manganese cells (3.9 Ω continuous load) at (a) 21°C and (b) −20°C. (By permission of Eveready.)

devices which have been designed to operate using 1.5 V aqueous cells – toys, games, cameras, portable cassette and CD players, radios, etc.

Soluble cathode systems

The principle of using a 'soluble depolarizer' is well established in aqueous cells. In the Grove cell of 1838, a zinc anode was coupled with nitric acid as cathodic reagent, using a platinum cathodic current collector. This system was later developed into the commercially successful Grove–Bunsen cell where a carbon current collector was substituted. In the Poggendorff or 'bichromate cell' of the same period, the catholyte was $Na_2Cr_2O_7$ in acid solution. The common feature of all successful cells of this type is a mechanism which prevents sustained attack on the anode by the cathodic reagent. In the case of lithium–organic cells, this mechanism derives from the formation of a passive layer on the metal, as described in Section 4.3 above.

All the currently available commercial lithium cells using a soluble cathodic reagent are based on sulphur dioxide. These advanced cells, which have outstanding performances but also certain drawbacks, have until fairly recently been limited to military uses. Now they are finding increasing service in a wider civilian market for heavy duty applications.

The cell may be represented as

$$Li(s)|SO_2, LiBr, AN|C(s)$$

with an overall reaction:

$$2Li(s) + 2SO_2(diss.) \rightarrow Li_2S_2O_4(s) \qquad (4.11)$$

An OCV of just over 3.0 V is observed. A typical electrolyte is a solution of LiBr in AN which has a conductance of 5 S/m at room temperature, falling to 2 S/m at $-50°C$. This rather small reduction leads to excellent low temperature discharge characteristics. The concentration of dissolved sulphur dioxide in an undischarged cell at room temperature gives rise to an internal pressure of about 300 kPa (3 atm). According to eq. (4.11), the internal pressure decreases as the cell is discharged. However, the pressure increases to 3 MPa (30 atm) at 100°C, as shown in Fig. 4.31, so that the engineering problems associated with designing and fabricating a safe container and seals are severe. The cells have a spiral wound construction as illustrated schematically in Fig. 4.32. The lithium foil strip is separated from the cathodic current collector, a carbon black/Teflon mix held on an expanded aluminium mesh, by a thin microporous polypropylene separator. A nickel-plated steel case encloses the spiral and acts as the negative terminal. The positive connector is attached to the cathode via a central tantalum post enclosed in a glass-to-metal seal. The can is hermetically sealed and the liquid phase is injected through a filling eyelet or 'fillpoint' which is then welded shut. A safety vent is located at the bottom of the cell

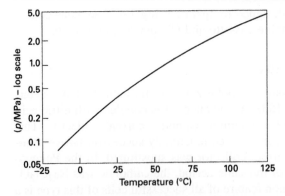

Fig. 4.31 Internal pressure of lithium–sulphur dioxide cells as a function of temperature. (By permission of Elsevier Sequoia; D. Linden and B. McDonald, *J. Power Sources*, 1980, **5**, 35.)

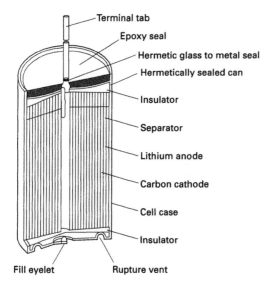

- Terminal tab
- Epoxy seal
- Hermetic glass to metal seal
- Hermetically sealed can
- Insulator
- Separator
- Lithium anode
- Carbon cathode
- Cell case
- Insulator

Fill eyelet   Rupture vent

Fig. 4.32 Lithium–sulphur dioxide spiral cell. (By permission of Elsevier Sequoia SA; as Fig. 4.31.)

to prevent rupture of the outer case if the internal pressure were to rise above the safety limit. Cells are fabricated using a balanced anode–cathode stoichiometric ratio.

Lithium–sulphur dioxide cells are characterized by energy densities of up to 300 Wh/kg (480 Wh/dm^3) which are at least three times as high as the best zinc and magnesium cell values. Because of the low internal resistance of the spiral wound cells, the discharge voltage of 2.7–2.9 V is hardly affected by the rate of discharge. In Fig. 4.33 typical discharge curves of a D-size cell (internal resistance 0.8 Ω) under various loads are shown. It can be seen that the discharge profiles are flat, with negligible voltage delay. Even at temperatures as low as −40°C, the voltage delay is minimal. The cell can operate at temperatures down to −54°C, where no conventional system is active, and up to +70°C. Cells of this size are capable of pulse loads as high as 30 A with a voltage no lower than 2 V. The shelf life of the cells is very long: it is claimed that less than 10% of the capacity is lost on storage over a 5 year period at room temperature.

At present, lithium–sulphur dioxide cells are manufactured in cylindrical form with capacities ranging from 1 to 30 Ah. They are classified as standard or high rate systems. The former are constructed to operate over a wide scale of load and temperature conditions. The latter are designed to deliver a high energy output at low operating temperatures. Cells are often produced as battery packs. Applications in the military area take advantage of the high power density and unrestrictive operating conditions, and include providing power for portable radio transceivers, night

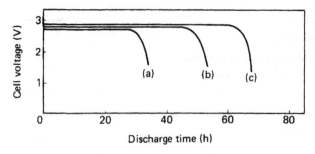

Fig. 4.33 Discharge curves for D-size lithium–sulphur dioxide cells at ambient temperature: (a) 270 mA; (b) 180 mA; (c) 140 mA. (By permission of Elsevier Sequoia SA: as Fig. 4.31.)

vision equipment, sonobuoys, missiles and 'artillery delivered' devices.* Emergency systems, alarms, aircraft emergency locators, etc. are typical examples of civilian uses. In 1995, following a 6 year flight, the NASA Galileo probe entered the atmosphere of Jupiter and sent telemetered data back to Earth via an orbiter spacecraft. The probe used three batteries, each comprising 13 lithium–sulphur dioxide cells which delivered 18 Ah, mainly in the last hour of the mission, and following deceleration forces of over 360 *g*.

Lithium–sulphur dioxide cells are characterized by high energy density, high power density, good voltage regulation, exceptional low temperature performance and superior shelf life. The real problem is that, if the cells are abused, they may explode or vent a highly toxic gas. Exposure of the cell to high temperatures, either external or internally generated by prolonged short circuit conditions, may raise the internal pressure to dangerous levels. Further, after storage at 60°C for extended periods, some corrosion of the glass-to-metal seal has been observed. Power sources based on the lithium–sulphur dioxide system must incorporate electrical protection devices such as current limiters or fuses, diodes to prevent charging on cell reversal, and thermal cut-off units. They must also have safety vents which allow controlled pressure release if the internal pressure exceeds the safety limits of the case design. As usage changes from predominantly military to civilian applications, even more stringent safety standards become essential. In particular, improvements in seal and vent technology must be made. A separate problem which must be solved before the use of this system becomes widespread concerns the safe disposal of spent cells.

Liquid cathodes

A number of inorganic molecules, such as thionyl chloride ($SOCl_2$), sulphuryl chloride (SO_2Cl_2) and phosphoryl chloride ($POCl_3$), have been found to be capable of acting both as solvent and as cathodic reagent in

---

* Tactical nuclear weapons are produced as 20 cm diameter shells.

Table 4.8 Properties of liquid cathode materials

Name	Formula	Relative molecular mass	m.p. (°C)	b.p. (°C)	Relative permittivity	Viscosity (Ns/m² × 1000)	Density (g/cm³)
Phosphoryl chloride	$POCl_3$	153.33	2	105	13.3 (22°C)	0.921 (25°C)	1.67 (25°C)
Sulphuryl chloride	SO_2Cl_2	134.97	−54.1	69.1	10 (22°C)	0.674 (25°C)	1.67 (25°C)
Thionyl chloride	$SOCl_2$	118.97	−105	78.8	9.25 (20°C)	0.603 (25°C)	1.65 (25°C)

lithium cells. Such materials are liquid over a wide temperature range and can dissolve considerable quantities of salts such as $LiAlCl_4$ to give conductive solutions. Moreover, they show excellent kinetic stability towards lithium, due to the formation of passivating layers, principally of lithium chloride. Liquid cathode cells have many similarities to the lithium–sulphur dioxide system discussed above, except that no organic component is present. From Table 4.8, which lists the physical properties of the three materials, it is seen that thionyl chloride has the widest liquid range: commercial interest has been focused mainly on this system.

The $Li–SOCl_2$ cell may be represented as

$$Li(s)|LiAlCl_4, SOCl_2(l)|C(s)$$

with the main cell reaction

$$4Li(s) + 2SOCl_2(l) \rightarrow 4LiCl(s) + SO_2(diss.) + S(s) \qquad (4.12)$$

The OCV is approximately 3.6 V. The generally accepted mechanism involves the initial formation of SO at the carbon cathode current collector. This molecule dimerises and finally decomposes to form sulphur and sulphur dioxide:

$$SOCl_2 + 2e \rightarrow SO + 2Cl^- \qquad (4.13)$$

$$2SO \rightarrow (SO)_2 \qquad (4.14)$$

$$(SO)_2 \rightarrow S + SO_2 \qquad (4.15)$$

The cell reaction thus leads to the formation of relatively insoluble lithium chloride at the cathode, which under some circumstances may block the porous carbon current collector, leading to an increased internal resistance. Another product is SO_2, which may lead to an increasing internal pressure as the discharge continues. However, in practice most of the SO_2 remains dissolved and the pressure build-up is small. Further electrochemical reduction of SO_2 to form lithium dithionite may also occur. A number of other possible reactions have been suggested, and as

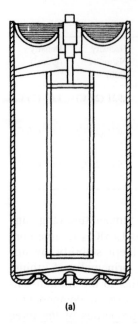   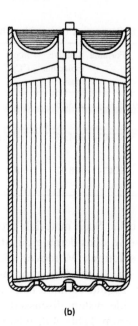

(a)   (b)

Fig. 4.34 Schematic structure of D-size lithium–thionyl chloride cells: (a) high energy type; (b) high power type

utilization of $SOCl_2$ is much higher at low current drains, it would appear that the overall reaction is a function of the discharge rate.

Practical cells are mainly of the cylindrical or button type. The former may be designed with slightly different internal structures for high energy or high power applications, as shown schematically in Fig. 4.34. The high energy cell is a bobbin-type unit, while the high power cell has a spiral construction giving a large electrode area. The cathode current collector is again a porous carbon black/Teflon mix, to which may be added small amounts of catalyst. Since $SOCl_2$ is a very aggressive chemical, there are special problems of material compatibility in cell fabrication. The cell case is constructed of nickel or nickel-plated stainless steel and must be hermetically sealed to a helium leakage rate of 10^{-9} cm^3/s or better. Normal plastics and rubbers are unstable in thionyl chloride, and only fully fluorinated polyolefins such as Teflon can be used as insulators. Early manufacturing problems concerned with attack on glass-to-metal seals have been solved. Much larger cells with capacities up to 10 000 Ah and 40 A discharge capabilities have been tested, and a 1.2 MWh 30 V battery has been developed for military use.

At room temperature, discharge curves are exceptionally flat (Fig. 4.35), even at rates of up to 3 A for D-sized cells. As noted above, $SOCl_2$ utilization is much higher at low current drains. Low rate cells manufactured by Mallory with a spiral configuration (4.45 cm $\times$ 25–38 cm electrodes) produced practical energy densities of 661 Wh/kg (1240 Wh/dm^3)

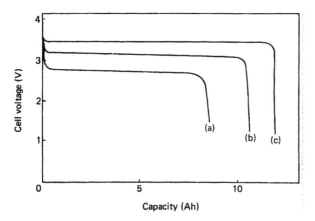

Fig. 4.35 Discharge curves of D-size lithium–thionyl chloride cells at ambient temperature: (a) 3.0 A; (b) 1.0 A; (c) 0.1 A

at 0.01 A. At temperatures of $-30°C$ and below (the working range for such cells is generally quoted at -40 to $+75°C$), some voltage delay is evident at higher current densities. Reasonably flat discharge curves are still developed for currents of 0.1 A and below (for D-sized cells), while 3 A can still be drawn without undue polarization, although the cell capacity is much reduced. Shelf life is excellent, with an estimated capacity loss of less than 0.5%/year over a 3 year period.

Low rate cells are considered to be as safe as conventional batteries and have been used in cardiac pacemakers. The safety of high rate Li–SOCl$_2$ cells, on the other hand, still presents serious problems, as there may be the danger of explosion for cells which have been short-circuited or reversed by forced discharge. However, according to one manufacturer, Tadiran, bobbin-type lithium-limited cells can be designed to withstand the most severe discharge/mechanical/heating tests. This is achieved by increasing the heat dissipation of the system by swaging the lithium foil to the inside wall of the casing, and by limiting the reactive electrode area. Low rate bobbin cells may not require the installation of pressure vents.

Many of the practical applications of lithium–thionyl chloride cells take advantage of their high energy density and favourable low temperature characteristics. Thus they are used in rocket launchers for space vehicles, balloon- and rocket-borne meteorological radiosondes, emergency locating transmitters, underwater instrumentation, etc. Cells are also manufactured with special dimensions and terminals for direct mounting on printed circuit boards. Miniature (button) cells based on the Li–SOCl$_2$ system have been used in a variety of implanted biotelemetry packages. The very large 40 A cells are used exclusively for military purposes, e.g. for driving torpedo motors. Heat management is a critical feature in the design of such batteries. Very large (e.g. 10 000 Ah) cells are used for standby power in missile silos – no maintenance or trickle charging is required.

Lithium–thionyl chloride cells are also constructed as reserve batteries

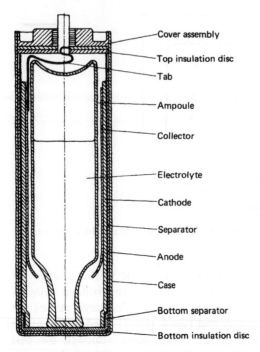

Cover assembly

Top insulation disc

Tab

Ampoule

Collector

Electrolyte

Cathode

Separator

Anode

Case

Bottom separator

Bottom insulation disc

Fig. 4.36 Cross-section of a lithium–thionyl chloride reserve cell. (By courtesy of Tadiran.)

for military use. A schematic diagram of the Tadiran system is shown in Fig. 4.36. The $LiAlCl_4/SOCl_2$ solution is contained in a sealed ampoule and the cell is activated by breaking this, either manually or by a spring-loaded or pyrotechnic device.

Sulphuryl chloride, when coupled with lithium, has a slightly higher energy density than thionyl chloride. The OCV of the cell

$$Li(s)|LiAlCl_4, SO_2Cl_2(l)|C(s)$$

is 3.91 V at 30°C. The cell reaction may be written:

$$2Li(s) + SO_2Cl_2(l) \rightarrow 2LiCl(s) + SO_2(diss.) \tag{4.16}$$

The higher OCV compensates for the increased mass of SO_2Cl_2 in comparison with SO_2Cl, so that sulphuryl chloride and thionyl chloride cells have comparable energy densities. A particular advantage of the SO_2Cl_2 cathode is that no sulphur is formed on discharge. Not only does this reduce the total amount of solid product and thus improve the efficiency of the carbon surface, but it also eliminates the possibility of direct reaction of sulphur with the lithium anode and the chance of thermal runaway. On the other hand, SO_2Cl_2-based cells may suffer from higher self-discharge and poorer performance at extremes of temperature. Bobbin-type cylindrical cells are manufactured, as well as large underwater batteries which use circulating electrolytes.

The energy density of liquid cathode lithium cells can be further enhanced to over 500 Wh/kg (1000 Wh/dm^3) by the use of halogen additives. BrCl, added to lithium–thionyl chloride cells, boosts the OCV to 3.9 V and prevents the formation of sulphur in the early stages of discharge. D-sized cells are manufactured. Addition of chlorine to lithium–sulphuryl chloride cells increases the energy density and improves the temperature-dependent electrical characteristics.

Phosphoryl chloride suffers from a relatively high melting point (2°C) and a consequently restricted liquid range and is not used in any commercial battery.

5 Secondary lead–acid cells

Kathryn R Bullock and Colin A Vincent

5.1 Introduction

By far the largest sector of the battery industry worldwide is based on the 'lead-acid' aqueous cell whose dominance is due to a combination of low cost, versatility and the excellent reversibility of the electrochemical system. Lead–acid cells have extensive use both as portable power sources for vehicle service and traction, and in stationary applications ranging from small emergency supplies to load levelling systems. In terms of sales, the lead–acid battery occupies over 50% of the entire primary and secondary market, with an estimated value of £100 billion per annum before retail mark-up.

5.2 The lead–acid cell

Introduction

The history of the lead–acid cell, more consistently named the 'lead–lead oxide cell', commenced in 1859 with the construction by the French physicist, Gaston Planté, of the first practical rechargeable cell, consisting of two coiled lead strips, separated by a linen cloth (Fig. 5.1). This system forms the basis of the most widely used secondary battery of the present time. Today lead–acid batteries are manufactured on every continent and in nearly every country. A total of about three hundred million lead–acid batteries are made every year. These range in size from 2 Wh cells to 100 Wh starting, lighting and ignition (SLI) systems, and to 40 MWh load levelling modules. The great success of this system is due to a number of favourable factors such as the relatively low cost and availability of the raw materials (lead and sulphur), ease of manufacture, long cycle life and

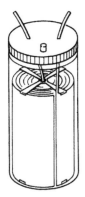

Fig. 5.1 Original Planté spiral wound lead–acid cell

favourable electrochemical characteristics. The performance of the lead–acid cell has been improved through more than a century of continuous research and development. This has been directed not only towards improved engineering design, materials and manufacturing techniques, but also to achieving a better understanding of the mechanisms of the cell process and of the factors which play an important role in degrading the system during its cycle life. The result has been a significant increase in the specific capacity, energy density and cycle efficiency, and a reduction in the need for cell maintenance – e.g. low rate charging to compensate for self-discharge.

The lead–acid cell can be represented schematically as having a negative electrode of porous lead (lead sponge) and a positive electrode of lead dioxide, PbO_2, both immersed in an aqueous solution of sulphuric acid:

$$Pb(s)|PbSO_4(s)|H_2SO_4(aq)|PbSO_4(s)|PbO_2(s)|Pb(s)$$

The overall electrochemical processes can be represented by the equation

$$Pb(s) + PbO_2(s) + 2H_2SO_4(aq) \underset{\text{charge}}{\overset{\text{discharge}}{\rightleftharpoons}} 2PbSO_4(s) + 2H_2O(l) \tag{5.1}$$

As the cell is discharged, sulphuric acid is consumed and water is formed. Consequently the electrolyte composition and density vary from about 40% by weight of H_2SO_4 (1.30 kg/dm^3) at full charge, with an associated OCV of 2.15 V at 25°C, to about 16% by weight of H_2SO_4 (1.10 kg/dm^3) when fully discharged, with an OCV of 1.98 V. The change in electrolyte specific gravity provides a convenient method of determining the state of charge of a cell. The open circuit voltage depends on the sulphuric acid (and water) activity and temperature and may be predicted with accuracy from thermodynamic free energy values. In Fig. 5.2, the OCV is given as a function of the percentage of discharge capacity. The discharge process also results in the formation of insoluble lead sulphate on both electrodes. This material is a very poor electrical conductor and its deposition in a

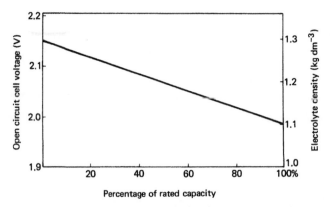

Fig. 5.2 Approximate open circuit voltage and electrolyte density as a function of percentage service capacity for the lead–acid cell

dense, fine-grained form can shield and passivate both electrodes, so that the practical capacity of a cell can become severely restricted – to as little as 5–10% of the theoretical capacity for large current densities. A number of techniques for reducing this effect will be described below. As discharge proceeds, the internal resistance of the cell rises, due to $PbSO_4$ formation and the decrease in electrolyte conductivity as H_2SO_4 is consumed.

During charge, $PbSO_4$ is reconverted to lead at the negative* and to PbO_2 at the positive.* The energy efficiency of the charge/discharge cycle may be high, but depends on charge rates and cell design.

In traditional 'wet' lead–acid cells the recharge reactions become less efficient as the cell approaches 85–90% state of charge, and the positive electrode begins to evolve oxygen. At the negative electrode, hydrogen ions react to form hydrogen gas, resulting in water loss. In some applications it is standard practice to supply at least 10% overcharge to the battery in order to obtain the full discharge capacity of the positive. This overcharge leads to further water loss from the electrolyte.

Valve regulated lead–acid (VRLA) batteries are designed to promote the chemical recombination of the oxygen at the negative electrode to minimize water loss. VRLA designs have the acid immobilized in a silica gel or absorbed in a porous glass separator with voids for oxygen transport. The oxygen diffuses from the positive to the negative plate where it oxidizes the lead, preventing it from reaching a potential where hydrogen will evolve. Since the plate is simultaneously on charge, the discharge product is immediately reduced to lead, restoring the chemical balance of the cell. The net sum of these chemical reactions is thus zero. Electrical energy input into the cell during charge is therefore converted to heat energy rather than chemical energy.

---

* 'Negative' and 'positive' are the almost universal terms used to identify the negative electrode and positive electrode, respectively, of a secondary cell (see p. 27).

5.3   Positive electrodes

The electrochemical reactions at the positive electrode are usually expressed as

$$PbO_2(s) + 4H^+(aq) + SO_4^{2-}(aq) + 2e \underset{\text{charge}}{\overset{\text{discharge}}{\rightleftharpoons}} PbSO_4(s) + 2H_2O(l) \tag{5.2}$$

In practice, the bisulphate ion, HSO_4^-, is a rather weak acid ($pK_a = 1.99$ at 25°C), so that for the sulphuric acid concentrations used in practical cells, the reactions

$$PbO_2(s) + 3H^+(aq) + HSO_4^-(aq) + 2e \underset{\text{charge}}{\overset{\text{discharge}}{\rightleftharpoons}} PbSO_4(s) + 2H_2O(l) \tag{5.2'}$$

may be considered a more accurate description of the electrode process. Both lead dioxide and lead sulphate are slightly soluble, and it is probable that soluble lead species are involved in the reaction mechanism. In the absence of mass transport limitations, it is considered that the process

$$PbO_2(H^+)_2 \text{ (surface)} + e \rightarrow Pb(OH)_2^+ \text{ (surface)} \tag{5.3}$$

is rate determining for the discharge reaction. The lead dioxide is found to vary slightly in stoichiometry at different stages in the cycle. Immediately after charge it may have a composition as high as $PbO_{2.05}$.

In order to obtain optimum current densities, it is necessary to use a highly porous structure so that the solid/electrolyte contact area is large. The fully charged positive electrode is therefore composed of a mass of small PbO_2 crystals connected to each other to form a continuous porous network. The electrode porosity is important for another reason, since it makes allowance for the increase in volume that occurs when PbO_2 is converted to $PbSO_4$. In a typical SLI battery the real area of the positive electrode is calculated to be 50–150 m^2 per Ah of capacity.

As current is drawn from the cell, the positive electrode voltage is depressed due to concentration polarization as sulphuric acid within the pores is consumed. This effect is more marked for partially discharged cells. Formation of lead sulphate decreases the pore volume and produces a microporous film on the PbO_2 matrix.

An important feature of the positive electrode discharge concerns the nature of the $PbSO_4$ deposit since the formation of dense, coherent layers can lead to rapid electrode passivation. Lead dioxide exists in two crystalline forms, rhombic (α-) and tetragonal (β-), both of which are present in freshly formed electrode structures. Since $PbSO_4$ and α-PbO_2 are isomorphic, crystals of lead dioxide of this modification tend to become rapidly covered and isolated by lead sulphate, and their utilization is less

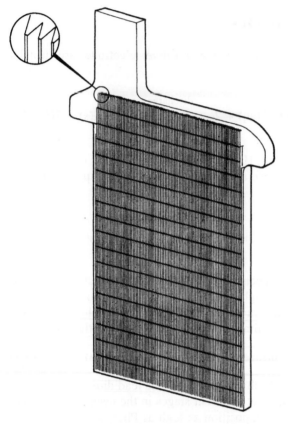

Fig. 5.3 Planté plate. Detail shows the grooves which increase the effective area of the plate

than that of the tetragonal β-form. As the latter is the thermodynamically more stable of the two, some transformation of α- into β-PbO$_2$ may occur during the life of a battery, with consequent improvement in its performance.

Positive electrodes are manufactured in three forms, as Planté plates, pasted plates and tubular plates. In Planté plates, the positive active material is formed by electrochemical oxidation of the surface of a cast sheet of pure lead to form a thin layer of PbO$_2$. The plate generally has a grooved surface to increase its surface above its geometric area by a factor of 3–10 (Fig. 5.3). Such plates have a very long life since they have a large excess of lead which can subsequently be oxidized to PbO$_2$. However, they are very heavy (and expensive) and their mechanical strength is poor, so that their use is confined to stationary battery applications in which long service life is important.

In 1880, Fauré proposed coating the lead sheet with a 'paste' of lead dioxide and sulphuric acid in order to increase the capacity of the system. It was soon found that the paste would be more readily applied to an open grid support, rather than to a lead sheet. However, the use of lead alloys

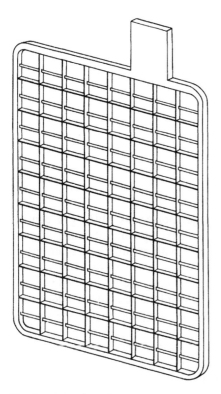

Fig. 5.4 Typical lead–acid battery grid: this acts as a framework to hold the active material in place

with superior mechanical properties to those of pure lead was required, although this resulted in other problems. Grid design (Fig. 5.4) is modified to suit a number of parameters (e.g. weight, corrosion resistance, strength and current distribution) which are important in different ways for different battery applications. Since the melting point of lead is low (327°C), most grids are formed by melting and casting; some lighter varieties are now manufactured using a stretching process which produces 'expanded metal' perforated sheets (Fig. 5.5); for very lightweight batteries, plastic lead composite grids are sometimes used. Grids are designed to ensure a low internal resistance for the cell and to minimize shedding of active material on cycling. Shedding causes loss of capacity, and dislodged material can accumulate on the battery case floor where it can give rise to short circuits between positive and negative plates. Additions of small quantities of tin are made to the lead to improve its coating properties, while antimony, calcium or selenium are added to form alloys with better stress resistance. Lead–antimony was the first alloy used, but lead–calcium alloys are becoming increasingly popular. Antimony (1.5–8%) greatly improves the mechanical properties of grids and connector bars, but also increases their electrical resistance, accelerates the self-discharge of the cell, and reduces

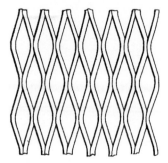

Fig. 5.5 Expanded metal grid

cycle efficiency. Further, during recharge, poisonous SbH_3 gas can be formed. Research is continuing into the development of better alloys. Use of lead–calcium or low antimony grids with as little as 1% antimony is increasing to reduce water loss. The main component of the paste used to fill the grid is known as lead dust, and consists of a carefully milled mixture of metallic lead and lead oxides. Water and sulphuric acid are added in a predetermined sequence, together with minor components and strengthening fibres, and the resulting slurry is loaded onto the grids and 'cured' or dried to produce a crack-free plate, with good adherence to the grid. The active material is later converted to the fully charged condition by the process of forming, as described below. Pasted plates have a relatively high capacity and power density, but are not mechanically strong. They are used extensively in SLI and similar batteries.

Tubular plates (also known as 'armoured' or 'clad' plates) consist of a row of tubes containing axial lead rods surrounded by active material (Fig. 5.6). The tubes are formed of fabrics such as terylene or glass fibre or of perforated synthetic insulators which are permeable to the electrolyte. They generally have a circular cross-section, but square, rectangular and oval tubes are also manufactured. Tubular plates are sufficiently strong to withstand continuous vibration and are resistant to shedding. They are also able to sustain many deep discharges without loss of integrity and are therefore suitable for applications such as EV traction.

5.4   Negative electrodes

The reactions of the negative electrode are generally given as

$$\text{Pb(s)} + \text{SO}_4^{2-}\text{(aq)} \; \underset{\text{charge}}{\overset{\text{discharge}}{\rightleftharpoons}} \; \text{PbSO}_4\text{(s)} + 2\text{e} \tag{5.4}$$

but as discussed previously in connection with eq. (5.2′), these are more correctly expressed as

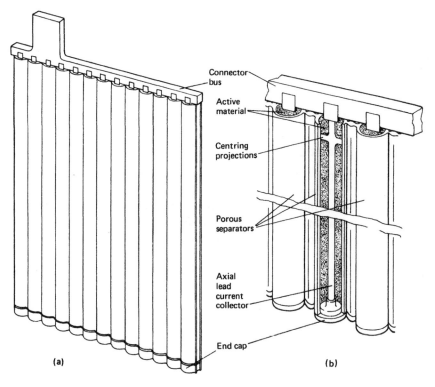

Fig. 5.6 (a) Tubular plates for lead–acid cells. (b) Cross-section showing central lead current collector, active material and porous separators

$$Pb(s) + HSO_4{}^-(aq) \underset{charge}{\overset{discharge}{\rightleftharpoons}} PbSO_4\ (s) + H^+\ (aq) + 2e \tag{5.4'}$$

Negative electrodes are almost exclusively formed of pasted plates, using either fine mesh grids or coarse grids covered with perforated lead foil (box plates) and the same paste used in positive plate manufacture. When the paste is reduced under carefully controlled conditions, highly porous sponge lead is formed consisting of a mass of acicular (needle-like) crystals which give a high electrode area and good electrolyte circulation. On deep cycling, however, and especially at high rates, the original morphology tends to alter to give larger crystal grains which have a lower overall area and which are more easily passivated by $PbSO_4$ layers. Additions ($\approx 1\%$) are therefore made to the negative mass to minimize crystallization of the spongy lead sulphate. Surface-active materials such as ligno-sulphonic acid derivatives (and substances deriving from soot or lampblack) are known as expanders and are thought to act by lowering the surface energy of the lead, so making the formation of large crystals less energetically favourable. They also affect the lead sulphate morphology. Additives such as very fine $BaSO_4$, which is isomorphic with $PbSO_4$, encourages the formation of a porous non-passivating layer of lead sulphate. The precise mechanism of the additive effects is complex and not completely understood. It is known that $BaSO_4$ and the organic additives interact, since together they are much more effective than the sum of their individual contributions.

5.5   Battery construction

Electrode forming

Forming is defined as the procedure undertaken, usually before final assembly in the battery case, to convert the active material in the positive and negative plates into their fully charged condition. In effect, the forming process is equivalent to the first charge, but the electrolysis operation is carried out using carefully controlled conditions of temperature and electrolyte composition, and low current density (typically < 100 A/m^2), in order to achieve complete conversion throughout the electrode masses while retaining adherence and porosity.

SLI batteries are also supplied in a dry charged state and are activated simply by filling with electrolyte. Plates for such batteries have extra additives, such as antioxidants in the negative active mass, and forming is followed by one of a number of controlled drying processes.

Separators

Separators are porous insulating sheets which are placed between the electrodes of opposite polarity to prevent contact and short circuit between them. They also act as a barrier to the transport of active material between the plates, mechanically support the positive active mass (so reducing shedding) and prevent dendrite formation. Effective separators must possess high porosity so that they will have low electrical resistance, low pore diameter in order to achieve good separation and, finally, resistance to oxidation and stability in highly acid conditions. In early batteries, separators were made from fir or cedar wood by removing the resinous substances in order to make them acid-resistant. Better results were obtained with microporous rubber separators: natural latex becomes highly porous after vulcanization due to water evaporation. However, present-day batteries almost all use separators made from synthetic polymers or glass filters. The porosity characteristics of these substances can be controlled to within close tolerances during the manufacturing process. The most popular polymer materials include sintered polyvinylchloride and extruded polyethylene. Special papers impregnated with phenolic resins are also in current use. Research is continuing to develop separators which are able to retain very large quantities of electrolyte for use in sealed batteries which can be operated in any orientation. Very efficient thin separators are required for high energy density batteries.

Hard rubber or glass fibre is used to fabricate retainers, which are perforated sheets in contact with the positive plate which protect the separators from its strong oxidizing environment.

Final assembly

Plates which have been tank-formed are first separated and cut to size. Lugs are milled free of oxide in preparation for welding to lead connector straps. The plates are assembled into parallel groups or stacks (usually with one extra negative plate), which are then interleaved, with separators, retainers and spacers inserted (Fig. 5.7). In a battery containing more than one cell group (Fig. 5.8), series connections must also be made. In large stationary batteries and older SLI models, these are made using external straps. In modern SLI batteries such connections are made within the battery case using through-partition ties, thus saving lead and reducing the internal resistance and weight of the battery.

Cases are now almost exclusively fabricated by injection moulding using synthetic polymers which have replaced the bitumen and hard rubber widely used in the past. Polypropylene has excellent mechanical and

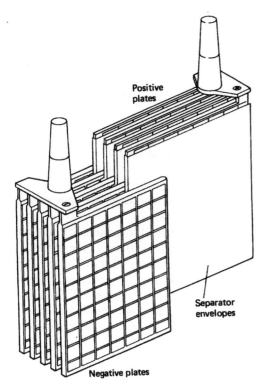

Positive plates

Separator envelopes

Negative plates

Fig. 5.7 Interleaving of positive and negative electrode groups (elements) to form a lead–acid cell

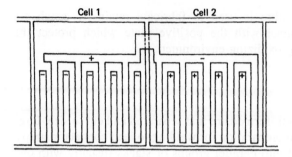

Fig. 5.8 Schematic diagram of a lead–acid battery showing through-partition connection

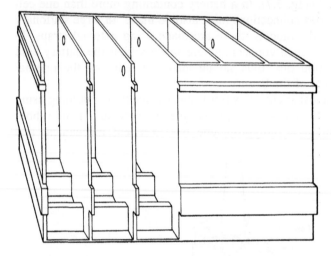

Fig. 5.9 Monobloc injection-moulded case

chemical properties and allows light, thin-walled monobloc containers to be constructed. Glass containers are still sometimes used for emergency batteries because they permit a rapid check of battery condition, but they are tending to be replaced by either glass–polyester materials with antiacid internal coating or transparent polyvinyl chloride.

Almost all lead–acid battery cases are rectangular (prismatic) in shape. The interior may be manufactured with projections to locate the plates and support them above the case floor, in order to leave mud or sludge space where shed materials can accumulate (Fig. 5.9). In other batteries the plates are suspended from the case lid or from a ledge in the case wall. The lid is welded or sealed to the case and is provided with apertures for terminal pillars, venting valves and simple screw caps or complex automatic systems for adding distilled water to the elctrolyte.

5.6   Performance and applications

Performance

Lead–acid batteries are designed for a wide range of applications, each of which has its own requirements and typical discharge pattern. For example, SLI batteries require short, very high rate discharges (at least 5 C) but are rarely discharged to any great depth. In contrast, EV traction batteries and some stationary industrial installations must be able to sustain deep discharges at effectively constant current (0.1–0.2 C, say). It is therefore not possible to discuss a generalized cycle performance for this system, and instead some details on the performance of certain important types are discussed below.

The practical capacity of all such batteries is also dependent on the temperature of operation and is found to drop very rapidly at temperatures below 0°C.

As the system is thermodynamically unstable with respect to hydrogen and oxygen evolution, lead–acid cells are subject to self-discharge:

$$PbO_2(s) + H^+(aq) + HSO_4^-(aq) \rightarrow PbSO_4(s) + H_2O(l) + \tfrac{1}{2}O_2(g) \quad (5.5)$$

and

$$Pb(s) + H^+(aq) + HSO_4^-(aq) \rightarrow PbSO_4(s) + H_2(g) \quad (5.6)$$

The rates of these processes are dependent on temperature, electrolyte volume and concentration, and most importantly, impurity content. If antimony is leached out of the positive grid it may be deposited on the negative plate where it catalyses reaction (5.6), because of its relatively low hydrogen overvoltage. The use of low antimony grids and antimony trapping separators reduces the amount of self-discharge. Reaction of the positive plate material with other solution impurities such as Fe^{2+} ions which can be re-reduced at the negative leads to very rapid self-discharge. To compensate for the loss in capacity due to self-discharge reactions, batteries may be placed on a maintenance charge when not in use.

Corrosion of the positive grid can occur on charging and overcharging if the metal becomes exposed to the electrolyte. This leads to a progressive weakening of the plate structure and to an increase in the internal resistance of the cell.

If a lead–acid battery is left for a prolonged period in an uncharged state or is operated at too high temperatures or with too high an acid concentration, the lead sulphate deposit is gradually transformed by recrystallization into a dense, coarse-grained form. This process is known as sulphation and leads to severe passivation, particularly of negative plates, and therefore inhibits charge acceptance. It is sometimes possible to restore a sulphated battery by slow charging in very dilute sulphuric acid.

Three main types of battery will be briefly described: SLI batteries, industrial batteries (traction and stationary) and small sealed portable batteries. The order follows their present relative commercial importance.

SLI batteries

As is well known, these batteries are used for cranking automobile internal combustion engines, and for supporting devices which require electrical energy when the engine is not running. The major part of the growth of the lead–acid battery industry in recent years is related to the world increase in the number of cars and delivery vehicles. About 80% of all lead–acid battery production goes to supply this market.

SLI batteries must be capable of supplying short but intense discharge currents at rates of over 5 C. They are therefore generally constructed of thin pasted plates, with thin composite separator/retainer layers and short connector buses to minimize the internal resistance. A typical battery is shown in Fig. 5.10. In this unit, the positive plates are inserted into pocket-shaped separators to increase their resistance to shock and prevent the shedding of material onto the cell floor. The through-partition connections reduce the internal resistance and weight of the battery. SLI batteries generally have nominal voltages of 12 V and 30–100 Ah capacities for cars, and 24 V and up to 600 Ah for delivery, construction and military vehicles. Typical batteries have energy densities of 30 Wh/kg (60 Wh/dm^3) but units with more than 40 Wh/kg (75 Wh/dm^3) may be obtained. Depending on use, service lifetimes of 3–5 years are normal.

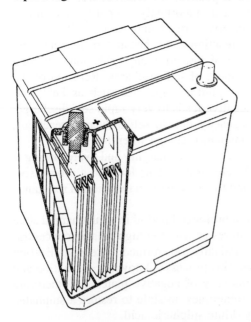

Fig. 5.10 Cut-away diagram of typical SLI battery. (By courtesy of Magneti Marelli.)

For vehicles used in rugged terrain, batteries with tubular positive plates are required. As discussed above, SLI batteries are sometimes supplied in a dry charged state, especially for markets where a long shelf life is needed.

Over the past 25 years, the introduction of batteries described as 'main-tenance-free' (MF) has had an important impact on the SLI market. What this term implies in practice is that no addition of water to the electrolyte is required over a normal service life of 2–5 years.

Standard lead–acid batteries lose a small amount of water by evapora-tion, but the major mechanism for water loss is by electrolysis to form hydrogen and oxygen, as described by eqs (5.5) and (5.6). The presence of small quantities of foreign elements lowers the overvoltages for these processes and leads to an accelerated water loss: it has been shown that the elements antimony, arsenic, cobalt, manganese, nickel, platinum and tellurium all have a deleterious effect, even at low levels. Two general approaches have been taken in the development of MF batteries:

- reduction of the rate of gas formation within the normal operating conditions of the battery;

- promotion of gas recombination.

Since SLI batteries are usually recharged at constant voltage, MF ver-sions must be constructed of such materials so that no substantial gassing occurs within the stabilized output voltage range of the alternator or dynamo. This implies that such batteries must contain modified positive grid and strap alloys in which the proportion of antimony is substantially reduced or eliminated and replaced by calcium or by a group of several metals (e.g. strontium/tin/aluminium). Calcium alloy grids have greatly improved behaviour so far as hydrogen evolution is concerned, but the calcium alloys have relatively poor processing characteristics, e.g. it is more difficult to fill complex grid moulds with such materials. MF batteries often have their electrolyte immobilized by means of gel formation using silica, calcium sulphate, etc., or by incorporating it in microporous glass separators. They are constructed in leak-proof enclosures with a one-way vent to release excess gas pressure and are sometimes referred to as 'semi-sealed' systems. These batteries have a much higher charge retention in comparison with more conventional units – up to 30% of capacity after 1 year. A further advantage is the absence of escaping acid vapours, so that corrosion of metallic components near the battery is eliminated. A 12 V maintenance-free semi-sealed truck battery is shown in Fig. 5.11.

Industrial batteries

Motive power batteries are generally of higher quality than SLI batteries. Their most important characteristics are constant output voltage, high volumetric capacity at relatively low unit cost, good resistance to vibration and a long service life. Since electric motors used for motive power require

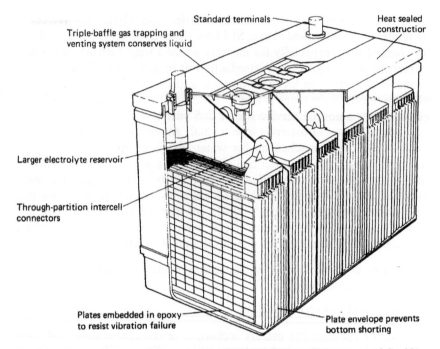

Standard terminals

Heat sealed
constructior

Triple-baffle gas trapping and
venting system conserves liquid

Larger electrolyte reservoir

Through-partition intercell
connectors

Plates embedded in epoxy
to resist vibration failure

Plate envelope prevents
bottom shorting

Fig. 5.11 12 V maintenance-free semi-sealed SLI truck battery. (By courtesy of Gould.)

high currents for long periods, traction batteries must be able to sustain prolonged and deep discharges followed by deep recharges, usually on a daily sequence. A typical discharge rate for this type of system might be C/5, and up to 80% of the nominal capacity would be required for daily service. The size and performance of traction batteries vary over a wide range: the voltage used may be 12–240 V and the capacity of each cell extends from 100 to 1500 Ah or more. The specific energy of these units is normally in the range 20–30 Wh/kg (55–77 Wh/dm^3) and the cycle life is 1000–1500 cycles. In Fig. 5.12, charge and discharge curves for a typical traction battery are shown. Prototype lead–acid traction batteries with energy densities of over 40 Wh/kg and 1000 cycle lifetime, and with energy densities up to 60 Wh/kg but with lower cycle lives, have been developed by a number of manufacturers. In Fig. 5.13 the assembly of the type of cell used in a traction battery is shown: the multitubular positive plates give the cell high specific energy and capacity and assure a long cycle life. Positive plates are sometimes constructed using pasted grids, but in this case it is necessary to incorporate glass wool felt and special separators to absorb shocks and vibration and to prevent shedding of active material. Traction batteries are used in industrial trucks, e.g. 'fork-lift' trucks, milk floats and other delivery vehicles, mining and other tractors, industrial sweepers and scrubbers, golf carts, etc. VRLA batteries are also under development for electric vehicles and hybrid vehicles.

Stationary batteries fall into two groups: systems for intermittent use

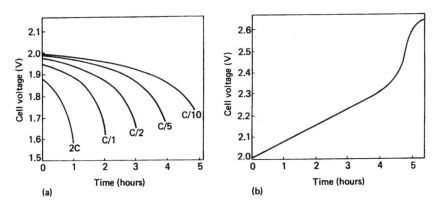

Fig. 5.12 (a) Discharge curves for a typical lead–acid cell at various rates. (b) Charging curve for the lead–acid cell at C/10

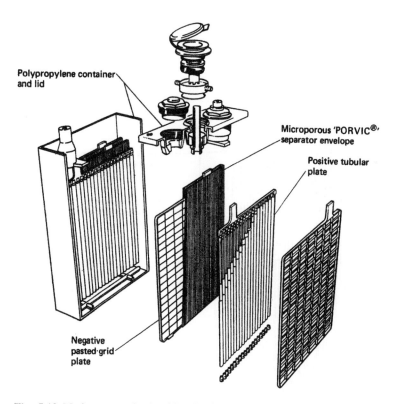

Fig. 5.13 Motive power lead–acid cell with tubular positive plates in which the active material is contained in pre-formed terylene tubes, and negative pasted grid plates surrounded by microporous polyvinyl chloride separator envelopes. The case and lid are formed of heat-sealed polypropylene. (By courtesy of Chloride Industrial Batteries.)

or standby power, and load levelling systems. The most important characteristics of the first group are reliability, long life and low self-discharge rate. Such batteries sometimes employ Planté plates; otherwise they use tubular plates or thick pasted plates formed on low antimony grids. An exploded view of a typical cell is shown in Fig. 5.14; the capacity of

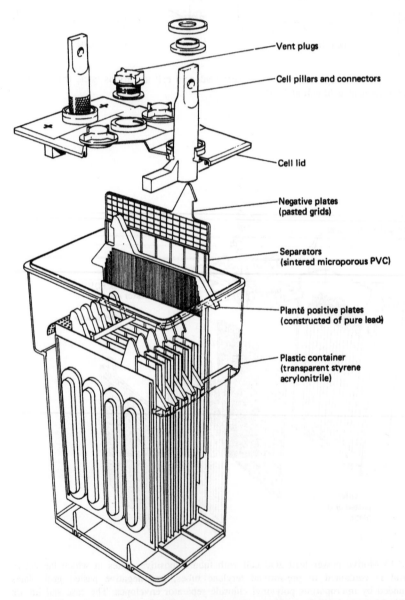

Vent plugs

Cell pillars and connectors

Cell lid

Negative plates
(pasted grids)

Separators
(sintered microporous PVC)

Planté positive plates
(constructed of pure lead)

Plastic container
(transparent styrene
acrylonitrile)

Fig. 5.14 Standby power high performance Planté cell. The container is fabricated from transparent styrene acrylonitrile which enables the electolyte level and cell condition to be easily monitored. (By courtesy of Chloride Industrial Batteries.)

each cell is in the range 20–2200 Ah at C/10. The energy density is about half that of traction batteries. Stationary batteries of this type are used for telecommunications systems, railroad signalling and track control and for standby power wherever it is necessary to have continuity of service during power cuts and momentary interruptions. Of the industrial markets, telecommunications and UPS have the highest growth rates. By 2001, standby batteries are expected to comprise 55% of total industrial lead–acid battery production.

Over the past few years a number of studies have been made of the use of lead–acid batteries for load levelling. The service required is very similar to that of traction batteries except that energy density is less important than cycle efficiency.

The number of VRLA batteries used in stationary applications is increasing rapidly. They account for more than 5.2% of the total US standby power production and more than 60% of Japanese and European production. Fig. 5.15 shows a VRLA battery design for telecommunications standby power. Over the past decade, VRLA batteries have been scaled to sizes up to 3000 Ah for industrial applications. Although the original

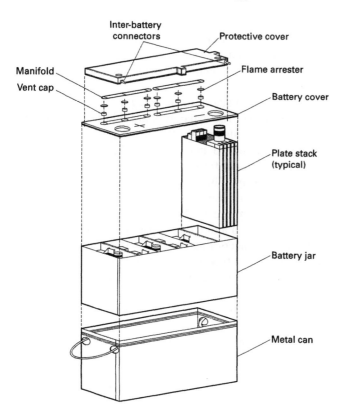

Fig. 5.15 Valve-regulated battery for standby power in telecommunications. (By courtesy of AT&T Laboratories Power Systems.)

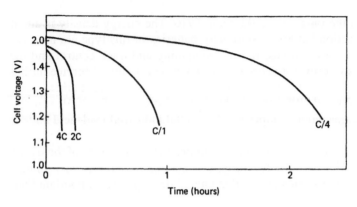

Fig. 5.16 Discharge curve for sealed cylindrical lead–acid cell

designs were made with pasted plates, commercial tubular designs are now available using gelled electrolyte.

Portable batteries

In recent years the market for small, portable lead–acid batteries has grown considerably. Almost all portable batteries are VRLA designs. Both cylindrical and rectangular (prismatic) unit cells are made and assembled into multicell packs.

Positive and negative plates are usually made using a 'honeycomb' grid support filled with active material. The separators are thin films of porous highly insulating materials which also retain the electrolyte. Usually they contain a non-woven glass microfibre mat which is heat and oxidation resistant. Cylindrical cells are made from plates and separators that are sandwiched together and rolled up. This assembly results in a vibration-resistant cell with low impedance, low polarization, long life and high utilization. A typical cell performance is shown in Fig. 5.16. These batteries are generally constructed with a range of capacities from 2 to 30 Ah; they give 300 to 2000 cycles according to the particular application. Further advantages offered by this type of cell include absence of maintenance, good constancy of the discharge voltage at up to C/4, ability to sustain short high current discharge pulses, low self-discharge (typically 6–8%/month) at ambient temperature and low cost in comparison with nickel–cadmium cells.

Portable valve-regulated lead–acid cells can operate in any orientation without acid leakage and find use in many different applications, such as in electronic cash registers, alarm systems, emergency lighting unit equipment, telephone boxes, switching stations, minicomputers and terminals, electronically controlled petrol pumps, cordless television sets and portable instruments and tools.

Recent prototype cylindrical cells with very thin plates can be discharged at a 20 C rate. This power capability is particularly important for power tools, e.g. electric drills. Thin, flat VRLA cells have also been developed for portable electronic markets. For example, a 2 V, 0.28 Ah cell with a thickness of only 6 mm is now available.

6 Secondary alkaline cells

Franco Bonino (in part) and Colin A Vincent

6.1 Introduction

The manufacture of secondary batteries based on aqueous electrolytes forms a major part of the world electrochemical industry. Of this sector, the lead–acid system (and in particular SLI power sources), as described in the last chapter, is by far the most important component, but secondary alkaline cells form a significant and distinct commercial market. They are more expensive, but are particularly suited for consumer products which have relatively low capacity requirements. They are also used where good low temperature characteristics, robustness and low maintenance are important, such as in aircraft applications. Until recently the secondary alkaline industry has been dominated by the cadmium–nickel oxide ('nickel–cadmium') cell, but two new systems are making major inroads, and may eventually displace the cadmium–nickel oxide cell – at least in the sealed cell market. These are the so-called 'nickel–metal hydride' cell and the rechargeable zinc–manganese dioxide cell. There are also a group of important but more specialized alkaline cell systems which are in use or are under further development for traction, submarine and other applications.

6.2 Cadmium–nickel oxide cells

Introduction

The first patent on an alkaline secondary battery was taken out by Waldemar Jungner of Sweden, who in 1899 proposed a system based on nickel hydroxide as the positive electroactive material, a mixture of cadmium and iron as the negative electrode, and an aqueous solution of potassium hydroxide as electrolyte. This cell, correctly called the cadmium–nickel oxide cell, is almost universally referred to as the 'nickel–cadmium' cell, and this latter term is used here. The nickel–cadmium system has been developed in different ways to produce a wide range of commercially important rechargeable systems including sealed maintenance-free cells with capacities of 10 mAh–15 Ah, vented standby power units with capacities of over 1000 Ah, cranking batteries capable of delivering peak currents of 8000 A, etc. Nickel–cadmium cells are characterized by long life, continuous overcharge capability, relatively high rates of discharge and charge, almost constant discharge voltage and the ability to operate at low temperatures. However, the cost of cadmium is several times that of lead, and this difference is unlikely to decrease since cadmium is a by-product of zinc production. Moreover, the cost of nickel–cadmium cell construction is generally more expensive than that of lead–acid cells, so that the overall capital cost of energy storage is up to 10 times higher. Health risks associated with the manipulation of cadmium and environmental concerns associated with cell disposal will undoubtedly affect future developments of this system. However, long cycle life, low maintenance and good reliability have made it an obvious choice for a number of applications such as emergency lighting, electricity grid switching operations, engine starting, etc., while the sealed system which can operate in any orientation over a wide temperature range has found use in 'cordless' electrical appliances of many types – power tools, portable television receivers, hedge trimmers, electric shavers, etc. The good low temperature performance has led to the wide use of nickel–cadmium batteries in aircraft and space satellite power systems. At the present time, nickel–cadmium batteries account for more than 7% of total battery sales in the Western world market.

The fully charged cell, which may be written as

$$Cd(s)|KOH(aq)|NiO(OH)(s)$$

has an OCV of 1.30 V at ambient temperature and the basic overall cell reactions are

$$Cd(s) + 2NiO(OH)(s) + 2H_2O(l) \underset{charge}{\overset{discharge}{\rightleftharpoons}} Cd(OH)_2(s) + 2Ni(OH)_2(s) \quad (6.1)$$

In practice, the cell reactions are more complex due to the formation of different NiO(OH) modifications and a series of higher nickel oxides with different degrees of hydration. The OCV of freshly charged cells may therefore be initially several hundreds of millivolts higher until these oxides spontaneously revert to NiO(OH) and oxygen. It is important to note that the electrolyte does not take part in the main cell reaction, so that its concentration is virtually independent of the state of charge of the cell. This is a particular advantage since neither the internal resistance nor the freezing point of the electrolyte are affected by the state of charge. (A minor disadvantage is that there is no simple way of assessing the state of charge, as can be done with the lead–acid battery by measuring the electrolyte density.) Depending on construction, nickel–cadmium cells have practical energy densities in the range 10–35 Wh/kg (30–80 Wh/dm^3) and a cycle life ranging from several hundreds for sealed cells to several thousands for vented cells.

Cell construction is mainly confined to two types, using either 'pocket plate' electrodes (vented cells) or 'sintered', 'bonded' or 'fibre plate' electrodes (vented and sealed cells). In the former, the active materials are retained within pockets of finely perforated nickel-plated sheet steel which are interlocked to form a plate. Positive and negative plates are then interleaved with insulating spacers placed between them. In sintered plate electrodes, a porous sintered nickel mass is formed and the active materials are distributed within the pores. In sintered plate vented cells, cellulose or other membrane materials are used in combination with a woven nylon separator. In sealed or 'recombining' cells, special nylon separators are used which permit rapid oxygen diffusion through the electrolyte layer.

Negative electrodes

The electrochemical reactions at the negative electrode are the comparatively straightforward processes:

$$Cd(s) + 2OH^-(aq) \underset{charge}{\overset{discharge}{\rightleftharpoons}} Cd(OH)_2(s) + 2e \qquad (6.2)$$

However, it is recognized that slightly soluble intermediates such as $CdO(OH)^-$ and $Cd(OH)_3^-$ are involved. Cadmium does not corrode since its equilibrium potential is more positive than that of hydrogen in the same solution. The active material in pocket plate cells consists of metallic cadmium, with up to 25% of iron and small quantities of nickel and graphite to prevent agglomeration. Two methods of preparation are used. One involves the electrochemical co-reduction of a solution of cadmium and iron sulphate; in the other, dry mixtures of cadmium oxide or hydroxide and Fe_3O_4 or iron powder are used. In some methods of pocket plate manufacture, the electrode material is pressed into pellets or 'briquettes' before being inserted into the pockets, and various waxes or oils may be used to facilitate this process.

Positive electrodes

The principal electrochemical reactions are represented by the following equation:

$$NiO(OH)(s) + H_2O(l) + e \underset{charge}{\overset{discharge}{\rightleftharpoons}} Ni(OH)_2(s) + OH^-(aq) \quad (6.3)$$

However, as pointed out above, some hydrated higher oxides of nickel (up to $NiO_{1.8}$) are initially formed on charge, which slowly decompose to give $NiO(OH)$ and oxygen. This situation is further complicated by the involvement of KOH (and possibly LiOH) from the electrolyte, and the formation of mixed alkali metal–nickel oxides. The net result of these additional reactions is an elevated OCV in freshly charged cells and some loss of cycle efficiency. Depending on charging rate and other factors, different crystal modifications of $NiO(OH)$ in addition to the predominant β-form may be produced, some of which cause considerable swelling and possible deterioration of contact between the active material and the current collector.

The active material used in pocket plate cells consists of $Ni(OH)_2$ together with up to 5% of $Co(OH)_2$, $Ba(OH)_2$, etc. to improve cell capacity and cycle life, and 20% of graphite in various forms to increase the electronic conductivity. The nickel hydroxide is precipitated from nickel sulphate in a controlled manner to produce fine particles of large surface area. As in the case of the cadmium electrode, the nickel hydroxide powder may be formed into pellets before insertion into the pockets.

Sintered, fibre and plastic bonded plates

Sintered electrodes were first produced in Germany about 60 years ago, but large-scale commercial manufacture did not start until after the Second World War. The preparation of sintered electrodes is more difficult and more costly than that of pocket plate electrodes, but leads to cells with superior electrical and mechanical characteristics. For high rate applications, accurate, thin section, closely spaced electrodes can be produced, and since the active material is in close contact with the current collector, the cell resistance remains fairly constant right up to the end of the discharge. The sintered plate substrates are made by pressing and then sintering at 800–1000°C in hydrogen nickel powder (usually formed by thermal decomposition of $Ni(CO)_4$) on a supporting nickel perforated foil or wire screen. By incorporating gas forming materials such as $(NH_4)_2CO_3$ in the nickel powder before sintering, a highly porous, mechanically stable matrix is produced. Up to 75–85% of the volume is void, and by carefully controlling the starting materials and thermal treatment, uniform pore radii ($\approx$5–30 μm) can be guaranteed. In the past 10 years or so, there have been significant developments to improve the porosity of the 'plaque' (while retaining mechanical strength) by forming powders of very low but uniform particle size with high surface area. An alternative approach is to deposit nickel from $Ni(CO)_4$ directly onto polyurethane foam or carbon-based felts. The thermal gas decomposition has excellent throwing power which permits an even distribution of nickel throughout fibrous matrix. Electrodes of this type are used for a wide range of high rate applications (including electric vehicles) and are sometimes referred to as *fibre plate systems*. Another technique is to form *plastic-bonded plates* with PTFE as a binder, using a pasting method and a nickel-plated steel mesh.

The sintered product is cut into the required shape and the active material is infused using one of a number of techniques – e.g. for the negative electrode by impregnating the sinter with concentrated aqueous cadmium nitrate, followed by thermal decomposition, or by cathodic polarization in molten $Cd(NO_3)_2$ baths, etc. The plates are washed and the impregnation cycle is repeated up to 5–10 times until the required loading is attained. Finally the plates are 'formed' by a sequence of carefully controlled charge/discharge cycles. Safety precautions are very important in the manufacture of cadmium-based electrodes because of the health hazards associated with this material.

Electrolyte

The electrolyte generally used is an aqueous solution of potassium hydroxide with a concentration of 20–28% by weight and a density of 1.18–1.27 g/cm^3 at 25°C. The low end of the range is used for cells designed to operate at normal temperatures (say above −10°C). About 1–2% of lithium hydroxide is usually added to such electrolytes to minimize coagulation of the NiO(OH) electrode on cycling. For low temperature applications, the more concentrated KOH solutions are used (without LiOH which increases electrolyte resistance); some problems with swelling are encountered with pocket plate electrodes and these concentrated solutions. Cells operating at high temperatures occasionally use aqueous sodium hydroxide as electrolyte.

Some water loss occurs, especially if there is prolonged overcharging of cells. Cells may be designed so as to contain a large reserve of electrolyte, so that topping up is only necessary after long intervals, e.g. 5–8 years in some stationary standby applications.

Vented pocket plate cells

The pocket plates are made from low carbon mild steel strip, 0.1 mm thick and perforated with 250–500 circular holes per square centimetre by needle or roller perforators. Each hole has a radius of about 0.1 mm. The strip is generally 16 mm wide and the outer edges are not perforated to facilitate later crimping operations. After perforation, the steel strip is nickel-plated and formed into channels by rollers as shown in Fig. 6.1. One channel is filled with active material while a second is fed in on top and crimped to close the pocket. The required number of filled strips are then interlocked together to form a battery plate. After guillotining to size, the cut edges are enclosed by welded U-section plated steel strip and current collector lugs are attached by spot welding. Finally the plates are pressed to form a flat rigid structure with a corrugated surface which permits insulating separator rods to be inserted (see Fig. 6.2).

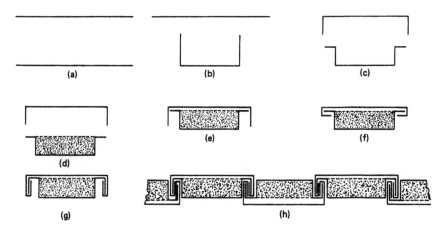

Fig. 6.1 Sequence of operations in fabrication of pocket plates for nickel–cadmium batteries. (a)–(c) Formation of channels in nickel-plated perforated steel strip. (d)–(f) Filling and crimping to form a long continuous pocket. (g)–(h) Interlacing of filled strips and compression to form final plate

Plates of the same polarity are then assembled with spacing washers on a collector bar to which is also attached the terminal pillar. Large assemblies are usually bolted together, while in small capacity systems the plates may be welded together. Double positive plates are used for heavy duty batteries with deep cycling regimes, while thinner plates are used in batteries designed for high rate/short discharge applications. Negative and positive electrode groups are interleaved as seen in Fig. 6.2, so that plates of opposite polarity are adjacent to one another and the thin insulating rods of polystyrene or similar material are inserted as shown. Plate separation varies from 1.0 to 3.5 mm. Steel containers are sometimes left with their bases open after assembly to permit the forming process to take place in large free electrolyte baths. Steel cases are generally electrically live, being connected to the positive electrode group.

Steel containers are used for applications involving mechanical stress, vibration, etc. However, corrosion protection measures must be carried out, such as nickel plating and external coating with epoxy resins. When a series combination of cells is required, they must be mounted in special wooden racks using locating pins to prevent short circuits. Translucent polystyrene or polyethylene containers are now widely used. Their freedom from corrosion, the ease of inspecting electrolyte levels and their light weight are obvious advantages. For stationary applications, batteries of various configurations can be formed simply by taping individual cells (in plastic containers) into blocks (Fig. 6.3).

Vented sintered plate cells

Sintered plate cells are assembled in a similar manner to pocket plate cells. Since the products of charging are formed within the pores of the rigid

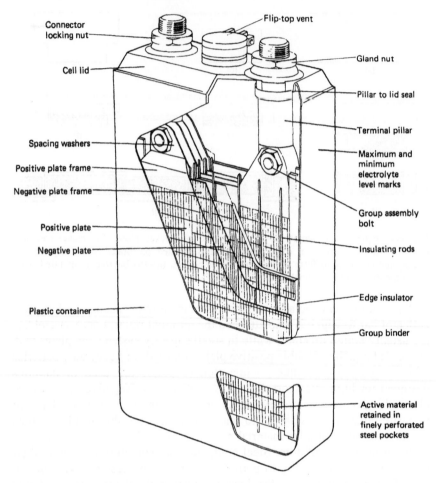

Fig. 6.2 Cut-away diagram of a typical nickel–cadmium cell. (By courtesy of Chloride Alcad.)

electrode matrix, there is no chance of swelling and the plates can be mounted closer together. Most manufacturers use woven nylon or other synthetic fibre separators; perforated polyvinylchloride sheet is also used.

Vented cells with tubular electrodes

Some versions of the cell are manufactured in which the active materials are supported in tubular elements made from spirally wound perforated steel ribbon. Such a design is more common for alkaline nickel–iron cells, and will therefore be described for the latter system.

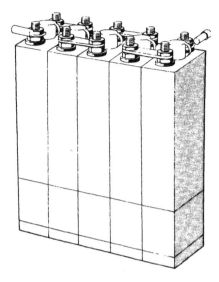

Fig. 6.3 Nominal 6 V battery formed by connecting five nickel–cadmium cells in series

Performance

Pocket plate cells have energy densities in the range 10–25 Wh/kg, depending on whether the cell has been designed for low or high rate duties. Sintered, fibre and plastic-bonded plate cells have a 50% higher energy density and can maintain higher rate discharges since they have lower internal resistance. Recently, increased power and energy densities have been achieved in 40 Ah aircraft batteries by optimizing design parameters and carefully choosing the separator material. Batteries of this type (24 V) can supply up to 23 kW of instant power at 25°C with a power density of 600 W/kg. At −30°C, the power is reduced by only 50%. Standing losses are lower for pocket plate than for sintered electrode cells. The inherent long life of the former (25–30 years in stationary batteries) is a notable advantage for emergency power applications.

Typical charge/discharge curves for nickel pocket plate cells at a number of rates are shown in Fig. 6.4. The relatively flat character of the discharge curves at up to the 5 hour rate is noteworthy. The distinct rise in the charging potential at about 90% of the capacity is due to the changeover from cadmium ion reduction to hydrogen evolution. Since some oxygen evolution occurs on the positive electrode during charge, it is necessary to supply up to 25% excess charge for sintered plate cells and 50% for pocket plate cells in order to achieve complete conversion of $Ni(OH)_2$; hence the cycle energy efficiencies of the two types of cell are 68% and 55% respectively. After an initial fairly rapid self-discharge (up to, say, 20%), self-discharge of nickel–cadmium cells is slow, except at elevated temperatures. In Fig. 6.5, charge retention is shown for a typical pocket plate cell as a function of time and temperature. It is seen that at 25°C,

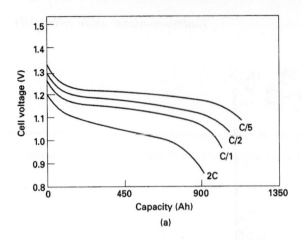

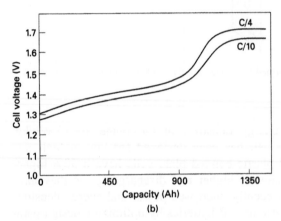

Fig. 6.4 Discharge (a) and charge (b) characteristics of a typical 900 Ah nickel–cadmium battery as a function of rate

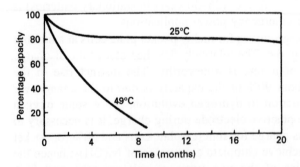

Fig. 6.5 Charge retention in nickel–cadmium cells after prolonged periods of open circuit. (By courtesy of Chloride Alcad.)

80% of the capacity is still available after 12 months. Self-discharge is higher with sintered plate electrodes. Unlike the lead–acid cell, nickel–cadmium cells can be stored for extended periods at any state of charge without damage.

Electrolytes may be contaminated by carbon dioxide over a number of years. When the carbonate concentration reaches a value of over 60 g/dm^3, the electrolyte must be replaced.

Sealed cells

Sealed nickel–cadmium secondary cells are designed so that no significant build-up of gas pressure occurs under normal working conditions, and since the electrolyte remains invariant, they require no maintenance. Edison patented an alkaline iron–nickel oxide sealed cell system in 1912, but commercial production of sealed nickel–cadmium cells began, initially in Europe, only 40 years ago. In a vented cell, overcharge causes oxygen evolution at the positive electrode and hydrogen at the negative electrode:

$$OH^-(aq) - e \rightarrow \tfrac{1}{2}H_2O(l) + \tfrac{1}{4}O_2(g) \qquad (6.4)$$

and

$$H_2O(l) + e \rightarrow OH^-(aq) + \tfrac{1}{2}H_2(g) \qquad (6.5)$$

On the other hand, if the cell is overdischarged (possibly due to other cells in the same battery having a slightly higher capacity), polarity is reversed and hydrogen is given off at the nickel electrode and oxygen at the cadmium. In sealed cells, protection against the effects of overcharge is brought about by incorporating excess cadmium hydroxide in the negative electrode. Then, when the positive electrode becomes fully charged, the negative electrode is still only partially charged. A continuation of the charging current thus results in oxygen evolution at the positive electrode (by the reaction of eq. 6.4), but further cadmium hydroxide reduction at the negative. The free oxygen can now diffuse to the negative electrode where it may be reduced electrochemically or react with cadmium. The cycle of oxygen evolution at the positive electrode and consumption at the negative electrode can continue indefinitely without affecting the cell. Cells can be fabricated which allow overcharge at C/10 without oxygen pressure ever exceeding 1 bar. Note also that the water content of the electrolyte remains constant. On terminating the overcharge current, oxygen pressure falls as the oxygen continues to be consumed by

$$\tfrac{1}{2}O_2(diss.) + Cd(s) + H_2O(l) \rightarrow Cd(OH)_2(s) \qquad (6.6)$$

Protection against overdischarge is achieved by the incorporation of a quantity of cadmium hydroxide – known as the 'antipolar mass' – in the positive electrode. Overdischarge then results in the formation of cadmium by reduction of the antipolar mass, rather than the evolution of hydrogen. Any oxygen evolved at the exhausted cadmium electrode will diffuse to the

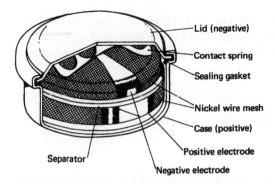

Lid (negative)
Contact spring
Sealing gasket
Nickel wire mesh
Case (positive)
Positive electrode
Separator
Negative electrode

Fig. 6.6 Sealed nickel–cadmium button cell. (By courtesy of Varta.)

positive electrode and combine with the cadmium according to eq. (6.6). It should be noted that the recombination reaction is exothermic, so that heat generation on overcharge can be a serious problem, especially with larger cells which, by their nature, have a relatively lower surface/volume ratio.

Sealed nickel–cadmium cells are manufactured as button, cylindrical and rectangular (prismatic) cells with nominal capacities ranging from 10 mAh to 15 Ah. The electrodes are commonly of the sintered type and their preparation is similar to that described above for vented cells. Some manufacturers incorporate polymeric bonding materials and use special pressing techniques. Button cells may also be assembled with circular pocket electrodes formed by pressing the active materials and containing the resulting pellets in fine nickel mesh. A separator disc of several layers of non-woven nylon or cellulose is inserted between the electrodes. Separator material and design is important in sealed cells, since oxygen flow between the electrodes must not be greatly impeded. A typical button cell of this type is shown in Fig. 6.6. Generally the positive electrode is placed at the bottom of the case in contact with an expanded metallic spacer. The negative electrode makes contact with the lid by means of a steel spring which serves to ensure good contact between the electrode masses and the electrolyte-containing separator. Button cells are used individually, or as series pack assemblies with different operating voltages, as illustrated in Fig. 6.7.

Special versions of these, usually with a nominal 3.6 V output and having appropriately sited terminals, are produced for mounting on printed circuit boards. Cylindrical cells, often manufactured as replacements for primary Leclanché cells of similar dimensions, are manufactured in two designs. In the most common version, a spiral construction is used, as shown in Fig. 6.8. The case is constructed of nickel-plated steel and acts as the negative terminal. The insulating disc at the bottom of the cell prevents short circuits between the rolled electrodes and the case. Note that the majority of cylindrical and rectangular cells have safety valves to release any excessive internal pressure resulting from, say, uncontrolled overcharging. Cylindrical cells with higher specific capacity but lower power have a

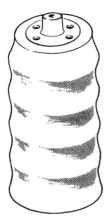

Fig. 6.7 Nominal 6 V sealed nickel–cadmium battery formed by connecting five cells in series. (By courtesy of Varta.)

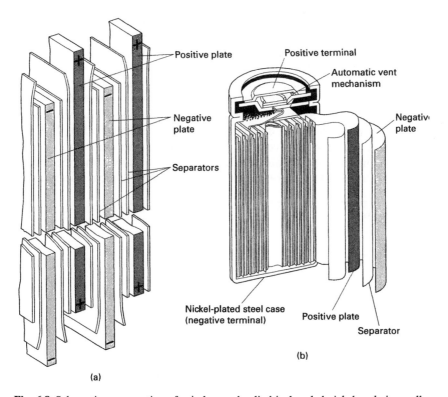

Fig. 6.8 Schematic cross-section of spiral wound cylindrical sealed nickel–cadmium cell

bobbin-type construction with the pressed positive electrode at the centre. Rectangular (prismatic) cells are fabricated from cut sintered positive and negative plates. Battery units of 6, 9 and 12 V are also manufactured using this design. Apart from the problem of heat dissipation, there is no reason why the sealed cell system could not be extended to larger pocket cells, and indeed some large sealed cell systems have been produced for emergency lighting applications where maintenance is particularly difficult.

Applications

Vented cells

Vented pocket plate cells have a wide range of applications that can be subdivided into three main groups. In the first, the cells are normally subjected to only shallow discharge, with occasional deep discharges, and are kept on a floating charge. The long life (> 30 years) and low maintenance requirement of nickel–cadmium cells under such a regime make them very suitable for emergency lighting, switch tripping in the electricity distribution industry, electric train control duties, etc. In the second group, cells are subjected to deep cycling at moderate rates (e.g. C/1–C/3) in such applications as train lighting, marine duties, telecommunications and traction (mine locomotives, industrial trucks, etc.). The third group comprises high rate cells with some deep cycling capability. Their main use, apart from short period emergency lighting, is for starting large diesel engines and gas turbines. A typical current demand curve for engine starting is shown in Fig. 6.9. In the initial 'breakaway' phase, currents of 1000–8000 A are required for periods of up to 1 s. Cells with large numbers of thin plates can deliver currents at rates greater than 10 C for this period, with the cell voltage dropping to no less than 0.65 V. The 'cranking current', usually about half that of the breakaway, may be needed for periods of 5–30 s before the engine starts, and a typical specification

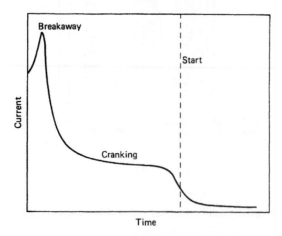

Fig. 6.9 Typical current drain during the starting sequence of large diesel engines or gas turbines. During the 'breakaway' phase, currents of 1000-8000 A are required for periods of up to 1 s. When the engine begins to turn, the current falls to a constant level until the engine fires

requires the starting sequence to be repeated six times. (Voltage recovery is almost instantaneous in nickel–cadmium cells, unlike the situation for the lead–acid system.) In most cases, except for very low temperature operations, the cranking current requirement determines the battery size.

Vented sintered plate cells are more expensive to produce than pocket plate cells and are therefore restricted mainly to applications where their mechanical integrity and resistance to shock and acceleration together with their high power density are of importance, e.g. in aircraft, helicopters, military vehicles, city buses, etc. Twenty-four volt batteries are generally used, with capacities in the range 30–60 Ah.

Advanced cells for EV applications

For the past 10 years, SAFT have been developing high performance 6 V modules for EV applications and have developed a large throughput automatic pilot line to provide power sources for Peugeot and Renault electric cars, with a capability of 200 000 modules per annum. The modules consist of five cells internally connected in series with capacities in the range 100–180 Ah. The energy density is in the range 53–63 Wh/kg and the power density is 171 W/kg. The expected cycle life on the vehicle is 2000. The cells are based on sintered positives and plastic-bonded negatives.

Sealed cells

Nickel–cadmium sealed cells are now a commercially important consumer product. They find use, both as button and cylindrical cells, in portable 'cordless' appliances such as power tools, electric razors and photoflash apparatus, and increasingly in 'hybrid' mains/battery equipment such as portable tape recorders, radios and television receivers. Many of these cells are readily interchangeable with primary batteries. In recent years, advances in design have increased recharge rates: cylindrical cells with sintered electrodes can now be fast charged from full discharge at up to the C/1 rate of 80% of capacity.

Sealed cells also have many important military and aerospace applications where absence of maintenance may be important. The battery for the 'Viking' Mars orbiting spacecraft, consisting of 26 sealed 30 Ah cells, is shown in Fig. 6.10.

Nominal 3.6 V batteries have been designed for direct mounting on printed circuit boards for CMOS and NMOS memory support applications. Such cells are normally float charged from the main microcomputer DC power supply and constitute a form of uninterruptible power supply (UPS).

'Memory effect' and capacity loss

The term 'memory effect' is used to describe a reversible process which results in the temporary reduction of the capacity of a nickel–cadmium cell

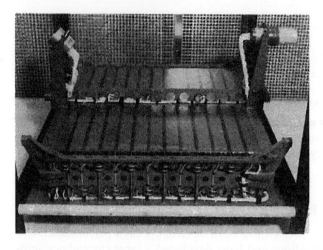

Fig. 6.10 Battery for 'Viking' Mars orbiting spacecraft, comprising 26 sealed 30 Ah nickel–cadmium cells, which was placed in Mars' orbit in 1976. (By courtesy of Jet Propulsion Laboratory.)

following repetitive shallow charge–discharge cycles. The effect may also be seen in some nickel–metal hydride cells.

If, following shallow cycling, a deep discharge is attempted, a step in the discharge curve is noted at the capacity value where the shallow cycle was reversed, i.e. the cell 'remembers' the depth of the shallow cycling. The size of the step (i.e. the voltage reduction) depends, *inter alia*, on the number of preceding shallow cycles and the value of the discharge current. The total capacity of the cell is not, however, affected since if the cell is now fully discharged (say to 0.9 V) and then recharged, a subsequent deep discharge will show a normal discharge curve, with no evidence of the step. What appears to be happening is that some morphological change occurs in the undischarged active material during the shallow cycling, which causes a reduction of the cell working voltage during subsequent discharge. It is not clear precisely what this change is, nor how it originates, although it has been suggested that the effect is based on an increase in the resistance of the undischarged material. It has very recently been suggested that the initial stage of the memory effect is caused by γ-NiOOH formation at the nickel electrode on overcharge during the repetitive shallow cycling.

'Memory effect' and 'voltage depression' are sometimes used interchangeably to describe the above phenomenon. A second effect, also commonly called 'voltage depression', is associated with overcharge of the negative (cadmium) electrode, especially at high temperatures, which results in nickel–cadmium alloy formation. The alloy discharges at a lower voltage than pure cadmium, but again a normal discharge behaviour can be restored by a full discharge–charge cycle.

Progressive irreversible capacity loss is sometimes confused with the reversible 'memory effect' processes. The former is a quite different phenomenon and is most commonly caused by a reduction in the electro-lyte volume due to evaporation at high temperatures through seals, or loss through pressure vents following overrated cycling or very prolonged overcharge. Permanent capacity loss may also be caused by degradation of the separator or by internal short circuits.

6.3   Metal hydride–nickel oxide cells

Introduction

The sealed nickel–metal hydride cell (more consistently metal hydride–nickel oxide cell) has a similar chemistry to the longer-established hydro-gen–nickel oxide cell considered in Chapter 9. In most respects (including OCV and performance characteristics), it is very similar to the sealed nickel–cadmium cell, but with hydrogen absorbed in a metal alloy as the active negative material in place of cadmium. The replacement of cadmium not only increases the energy density, but also produces a more environ-mentally friendly power source with less severe disposal problems. The nickel–metal hydride cell, however, has lower rate capability, poorer charge retention and is less tolerant of overcharge than the nickel–cadmium cell.

Hydrogen absorption alloys

Hydrogen absorption alloys were discovered in the 1960s during research into magnetic materials. They can absorb over a thousand times their own volume of hydrogen and usually consist of two metals, one of which on its own absorbs hydrogen exothermically and a second for which the process is endothermic and which serves as a catalyst for the dissociative adsorp-tion which precedes diffusion of atomic hydrogen into the lattice. The most important are the AB_2 series (e.g. $ZrNi_2$) and the AB_5 series (e.g. $LaNi_5$). Partial substitution of one or other of the two components allows the alloy to be 'fine-tuned', e.g. to control volume expansion, surface film formation, etc., and hence to provide an optimized negative electrode. Lanthanum is generally substituted by the less expensive naturally occurring *misch metal* which contains other lanthanides such as Ce, Pr and Nd. Alloys with up to eight metallic components have been described. Each alloy has a charac-teristic equilibrium hydrogen pressure at a given temperature, although in practice there is always a slight hysteresis in the absorption/desorption cycle. In addition to application in electrochemical cells, these alloys are used for hydrogen storage/transportation and have been considered for use in heat pumps, internal combustion engines and hydrogen purification.

Cell reaction

In an electrochemical cell, the reaction at the negative may be written as

$$M(s) + H_2O(l) + e \underset{\text{discharge}}{\overset{\text{charge}}{\rightleftharpoons}} MH(s) + OH^-(aq) \qquad (6.7)$$

where the metal hydride is in (or near) equilibrium with gaseous hydrogen and so has an emf given by

$$-\frac{RT}{2F} \ln p_{H_2}$$

The reaction at the positive is the same as that in the nickel–cadmium cell:

$$NiO(OH)(s) + H_2O(l) + e \underset{\text{charge}}{\overset{\text{discharge}}{\rightleftharpoons}} Ni(OH)_2(s) + OH^-(aq) \quad (6.8)$$

so that the overall cell reaction is

$$M(s) + Ni(OH)_2(s) \underset{\text{discharge}}{\overset{\text{charge}}{\rightleftharpoons}} MH(s) + NiO(OH)(s) \qquad (6.9)$$

The cell is written as

$$MH(s)|KOH(aq)|NiO(OH)(s)$$

As in the nickel–cadmium cell, the electrolyte is concentrated potassium hydroxide. Depending on the metal alloy used, the emf has a value usually in the range 1.32–1.35 V, which turns out to be almost the same as that of the nickel–cadmium cell. Note that the electrolyte composition is completely invariant during cycling. Unlike the situation with the nickel–cadmium cell, water is not involved in the cell reaction.

The balance of the cell is arranged so that it is limited by the nickel hydroxide. Hence oxygen is evolved at the positive rather than hydrogen at the negative on overcharge. Provided that the current towards the end of the charging cycle or on overcharge is limited, the oxygen can diffuse through the separator to recombine according to

$$4MH(s) + O_2(aq) \rightarrow 4M(s) + 2H_2O(l) \qquad (6.10)$$

and thus prevent any build-up of pressure. The negative electrode has also discharge reserve incorporated.

Cell construction

Button, cylindrical and prismatic sealed cells are similar in design to the starved-electrolyte configuration of nickel–cadmium cells. A schematic diagram of a six-cell battery is shown in Fig. 6.11. Because of the slightly

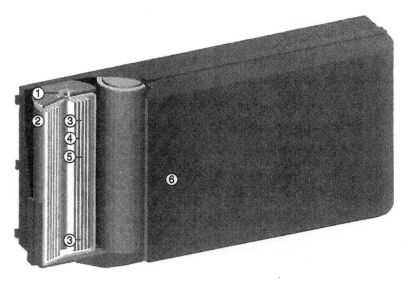

Fig. 6.11 Six-cell nickel–metal hydride battery. 1, positive cap, connected to the nickel oxide electrode; 2, can, connected to metal hydride electrode and serving as negative terminal; 3, separator; 4, cathode; 5, anode; 6, plastic battery case which contains interconnected cells and electronic management system. (By permission of Duracell.)

different operating conditions (e.g. completely invariant electrolyte concentration), the positive electrode is formulated in a slightly different way in order to optimize performance. The negative may be surface-treated to facilitate hydrogen absorption and discourage gassing at high charge rates. Hydrophilic polypropylene separators have been developed to improve the poor self-discharge characteristics of this system. Nickel–metal hydride cells incorporate resealable safety vents.

Performance

Nickel–metal hydride cells can be discharged at the 2 C rate (and in some cases at 4 C) and charged at 1 C. An AA-sized cell with a nominal capacity of over 1 Ah can thus be discharged at over 2 A and with a peak current of over 10 A. The energy density is highly dependent on rate, but for comparable conditions is 25% higher than an equivalent nickel–cadmium cell. Fig. 6.12 shows a comparison of the discharge characteristics of these two systems.

Self-discharge is probably the principal drawback of the nickel–metal hydride cell, with values of up to 4–5% per day being reported. This is caused principally by hydrogen dissolved in the electrolyte (i.e. in equilibrium with hydrogen absorbed in the alloy) reacting with the positive electrode. Such cells may also suffer from voltage depression or the memory effect in which the active material not utilized in shallow cycling undergoes some change of state, which renders it unable to take part in the discharge until a full discharge–charge cycle has been carried out.

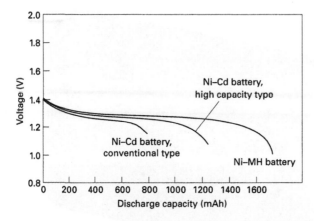

Fig. 6.12 Comparison of discharge characteristics for Ni–Cd and Ni–MH cylindrical cells at 20°C. Cell height = 42 mm, cell diameter = 16.5 mm. Charge = 1 C, discharge = 0.2 C. (By permission of National/Panasonic.)

Cells are usually recharged at constant current at a rate which is limited by the need to control the cell temperature and ensure that oxygen produced at the positive towards the end of the charging cycle, or on overcharge, can diffuse to the negative. Various methods of charge control may be used: the most popular, as with nickel–cadmium, is known as the '−ΔV method' and relies on detecting a small drop in the charging voltage which occurs when the cell is fully charged.

Applications

In 1990, Sanyo and Matsushita initiated large-scale commercialization of small sealed nickel–metal hydride batteries. They are now joined by Duracell, Toshiba and Varta in a consortium which is known as the '3C alliance' (camcorders, cellular telephones and computers). Several plants have been commissioned which are each producing 100–200 million cells per annum. It is forecast that nickel–metal hydride may overtake nickel–cadmium before the end of the century. In addition to the '3Cs', nickel–metal hydride cells are used for a wide variety of cordless consumer products, communications equipment and other high rate long cycle life applications.

This cell system is also under active development for EV traction. Realistic targets are reported as 80 Wh/kg for energy density, 200 W/kg for power density, and a cycle life of over 1000. The Ovonic Battery Co have already demonstrated 67 Wh/kg for a 250 Ah battery, and Panasonic have reported a similar value for a 130 Ah unit. Thermal management and cost are likely to be the key factors limiting development of nickel–metal hydride technology for electric vehicles.

6.4 Zinc–manganese dioxide cells

Introduction

The idea of a rechargeable zinc–manganese dioxide cell has been under consideration for very many years. Attempts to recharge Leclanché or zinc–carbon primaries were always doomed to failure because of the innate irreversibility of the discharge reactions (considered in Chapter 3). With the introduction in the 1960s of the alkaline manganese primary with its simpler chemistry, it became feasible to produce a practical secondary system, provided that the cathode reduction was limited to a nominal value of $MnO_{1.5}$, since the step in which Mn III is reduced to Mn II produced cathode expansion and soluble species, and is essentially irreversible. A second problem which had to be addressed is dendritic growth of zinc during charge which can lead to internal short circuits. In the 1970s, Union Carbide developed commercial secondary cells and batteries. Initial discharge characteristics (to 0.9 V) were similar to those of the primary system, but the available energy and power declined rapidly as the cells were cycled. Overcharge was also a problem, as was cell failure or leakage following deep discharge. Despite the low cost of these cells, they had little commercial success.

In the late 1970s and 1980s, a major research programme headed by Kordesch at the Technical University Gratz, latterly in collaboration with Battery Technologies Inc., transformed the prospects of what have become known as RAM cells. These now have greatly improved capacity and reliable cycling behaviour. Mercury has been eliminated from the anode while good shelf life at elevated temperatures has been retained. The technology has been licensed to a number of companies, the major manufacturer currently being Rayovac who began mass production in 1993, and who market cells as RENEWAL® Reusable Alkaline™ batteries. Other licensees include the Pure Energy Battery Corporation in Canada and the Young Poong Corporation in Korea (ALCAVA™).

The cell may be written as

$$Zn(s)|KOH(aq)|MnO_2(s),C(s)$$

and the cell reaction as

$$Zn(s) + MnO_2(s) + \tfrac{1}{2}H_2O(l) \rightarrow MnO.OH(s) + ZnO(s) \qquad (6.11)$$

The cell emf at 25°C is 1.55 V. As noted in Chapter 3 where the primary cell was described, the discharge proceeds at the cathode by the movement of protons and electrons into the MnO_2 lattice. Provided that the discharge is not allowed to proceed further than the one electron level – for example, by using a zinc-limited anode or a 0.9 V cut-off voltage – the process can be reversed during charge. It is found, however, that the capacity of the cell falls with cycle number, and that the number of useful cycles that may be obtained depends markedly on the depth of discharge.

Cell design

The cell design is similar to the inside-out construction of the primary cell (Fig. 6.13). The cathode uses EMD (electrolytic manganese dioxide) with 10% graphite and is formed into four pressed ring-shaped pellets. It may include small quantities of additives such as catalysts for recombining hydrogen and inert powders to control the porosity. It has been shown that it is important to optimize both the electronic and ionic conductivities in the cathode if good MnO_2 utilization is to be achieved. The anode consists of zinc powder in a gelled KOH matrix, with organic inhibitors to reduce corrosion and a central metallic current collector. A high concentration of KOH is required for effective cell operation since zincate is initially formed by the reaction

$$Zn(s) + 4OH^-(aq) - 2e \rightarrow Zn(OH)_4^{2-}(aq) \qquad (6.12)$$

followed by the formation of a porous layer of zinc oxide around the zinc particles:

$$Zn(OH)_4^{2-}(aq) \rightarrow ZnO(s) + 2OH^-(aq) + H_2O(l) \qquad (6.13)$$

ZnO is added initially to the KOH electrolyte to reduce the possibility of hydrogen formation on charge. It is very important to prevent the discharge proceeding beyond the first electron, which in practice means before $MnO_{1.6}$. At voltages lower than 0.9 V, soluble species are generated which give rise to irreversible behaviour and zinc corrosion. Attempts have been made to increase the capacity by using the second electron in positives having large quantities of active carbon, but the low voltage of the discharge makes the energy gain rather small. Commercial cells use zinc limitation to control the MnO_2 discharge.

A two- or multilayer separator is generally used in which a strong fibrous element is incorporated to prevent internal short circuits by zinc dendrites

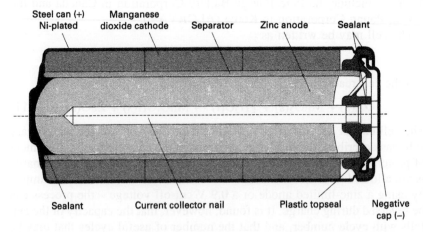

Steel can (+)   Manganese
Ni-plated   dioxide cathode   Separator   Zinc anode   Sealant

Sealant   Current collector nail   Plastic topseal   Negative cap (−)

Fig. 6.13 Cross-section of a RAM cell. (By permission of Professor K. Kordesch.)

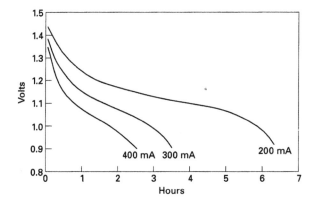

Fig. 6.14 Voltage profile for constant current discharge of a RENEWAL® AA RAM cell. (By permission of Rayovac.)

formed during charge. Cells have standard crimp seals and safety vents. Rayovac have designed an external plastic sleeve which, unlike those used in primary cells, has no 'overwrap' at the top of the cell case. This may be exploited by the design of 'discriminating' charge/discharge external contacts, i.e. to prevent the charging of primary cells in chargers designed for RAM cells.

Cells are manufactured in AAA, AA, C and D sizes, with the AA cell being the most common. The demand for AAA cells is now increasing rapidly.

Performance

RAM cells are manufactured and shipped charged and have an initial capacity of about 1.8 Ah for AA-sized cells discharged at 50 mA (in comparison with, say, 2 Ah for an equivalent primary cell). This capacity falls to 1 Ah after storage for 3 years at room temperature. At higher drains, the initial capacity drops to about 0.6 Ah at 400 mA (Fig. 6.14). Cells are designed to operate within a temperature range of 0–65°C. The higher internal resistance of RAM cells limits their maximum continuous output current and also their peak output currents in comparison both with primary cells and with nickel–cadmium and nickel–metal hydride secondary cells. A new cell will have an internal resistance of approximately 0.1 Ω, but this will rise to 0.25 Ω with use.

Cycling RAM cells produces significant progressive capacity loss, especially for deep discharge as shown in Figs 6.15 and 6.16. As a rule of thumb, the capacity of the 25th cycle is approximately half that of the initial cycle. A typical graph of capacity fade is shown in Fig. 6.17. The degree of capacity fade is reduced if the discharge is terminated at higher voltages, corresponding to lower capacity usage. Under partial discharge conditions, RAM cells can provide many hundreds of cycles with little

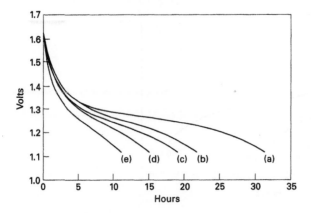

Fig. 6.15 Effect of charge–discharge cycles on discharge curve of a RENEWAL® AA RAM cell. Constant current discharge at 50 mA. (a) Cycle 1, (b) cycle 5, (c) cycle 10, (d) cycle 25, (e) cycle 50. (By permission of Rayovac.)

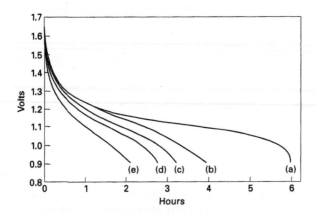

Fig. 6.16 Effect of charge–discharge cycles on discharge curve of a RENEWAL® AA RAM cell. Constant current discharge at 200 mA. (a) Cycle 1, (b) cycle 10, (c) cycle 25, (d) cycle 50, (e) cycle 100. (By permission of Rayovac.)

capacity fade (Fig. 6.18). It is important to note that these cells do not suffer from memory effect, and can supply the normal deep discharge capacity after multiple short discharge–charge cycles. If RAM cells are to be used for higher rate applications, it has been shown that it is preferable to use parallel arrays of AA cells, rather than C- or D-sized cells.

The self-discharge rate of RAM cells is approximately 0.01% per day, which gives them a clear superiority over nickel–cadmium and nickel–metal hydride cells (Fig. 6.19).

Constant potential charging or constant current charging with a voltage cut-off at 1.65 V to prevent the formation of soluble Mn (VI) species may be used. Specially designed chargers, which permit a rapid, optimized cycle, use 50–60 Hz current pulses and monitor the OCV during the current

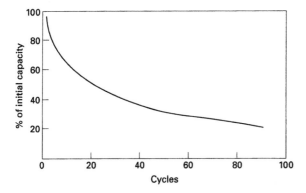

Fig. 6.17 Loss in capacity as a function of cycle number following full discharge for a RENEWAL® AA RAM cell. (By permission of Rayovac.)

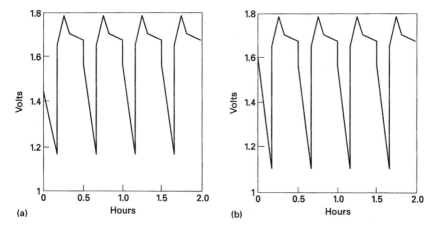

Fig. 6.18 Repetitive short discharge cycles for a RENEWAL® AA RAM cell; 400 mA constant current discharge for 10 minutes followed by standard recharge. (a) Cycles 1–4, (b) cycles 707–800. (By permission of Rayovac.)

interruption. Depending on the measured voltage, the charger blocks a proportion of the pulses, such that the current tapers to zero towards the end of the charging process. Because of the possibility of cell imbalance, series-connected batteries cannot be charged without taking special precautions. This contrasts with nickel–cadmium cells where limited continuous overcharge can be tolerated. There is, however, a straightforward electronic method for dealing with this problem, using a low cost overflow diode which starts to conduct at 1.6 V and carries all of the current at 1.7 V. Faster charging is also possible since any 'overcharge' heat is evolved outside the cell in the diode . To prevent series-connected cells reversing on overdischarge, reversal limitation diodes operating at −0.8 V are used. While it is usually impractical to use solar charging with nickel–cadmium

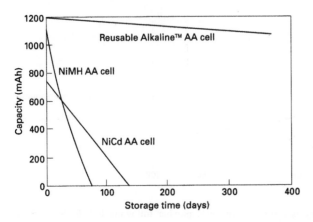

Fig. 6.19 Comparison of charge retention on storage of a RENEWAL® AA RAM cell with typical nickel–cadmium and nickel–metal hydride cells. (By permission of Rayovac.)

or nickel–metal hydride cells, because of the level of self-discharge at high temperature, this is not the case for RAM cells.

A comparison of Rayovac's RENEWAL® cells with typical nickel–cadmium and nickel–metal hydride cells is given in Table 6.1. The advantages of RAM technology are low cost, absence of heavy metals, excellent shelf life and capacity retention, and lack of memory effect. The disadvantages are clearly their limited cycle life and the decrease of capacity on deep discharge. These cells are inappropriate for high current applications where frequent deep cycles are required, as in power tools or laptop computers. They are also incapable of being used in the extreme conditions that are possible with lithium ion cells (Chapter 7), nor do they have comparable energy density of the latter. They are, however, likely to prove strong competition for application in devices such as mobile/cordless telephones, personal audio, electronic organizers, cameras, toys and games, because of their low cost, and have already captured more than 50% of the direct consumer market (i.e. principally supermarket outlets) in the USA.

The commercialization of RAM cells has presented a rather novel type of power source to the consumer market. First, they are ready to use as purchased and retain their charge well until used or after charging (Fig. 6.19), and thus behave rather like a primary battery in this respect. Second, they are not much more expensive than an alkaline manganese primary and thus are very cost-effective, even if capacity fade limits their use to 20 or so cycles. It will be interesting to discover how deeply this new technology penetrates the primary cell market in the near future.

Table 6.1 Comparison of AA-cell characteristics (by permission of Rayovac)

	Nickel–cadmium	Nickel–metal hydride	RAM
Nominal capacity (mAh)	750	1100	1400 (initial)
Cycle life	200+	300+	25+
Working voltage (V)	1.3–1.0	1.3–1.0	1.4–0.9
Weight (g)	22	26	22
Energy density (Wh/kg)	41	51	80 (initial)
Energy density (Wh/dm^3)	115	170	220 (initial)
Maximum rated current (A)	>5	>2	0.5
Peak current (A)	>10	>10	1
Self-discharge rate at room temperature (% per day)	1	4	0.01

6.5   Iron–nickel oxide cells

Introduction

The iron–nickel oxide alkaline battery system has many features in common with the nickel–cadmium system discussed above. It was first developed by Edison in the USA at the turn of the century and was patented in the same year as Jungner's first nickel–cadmium US patent, 1901. Iron can be regarded as a favourable active battery material because of its low cost, high theoretical specific capacity (twice that of cadmium) and non-toxic, pollution-free characteristics. However, because its reduction potential is below that of hydrogen, and since hydrogen overvoltage is low on iron, charge retention is poor and efficiency is low.

Manufacture of iron–nickel oxide batteries commenced in 1908, but the system did not have the commercial success of nickel–cadmium. Until comparatively recently, there was only a very limited production of stationary batteries in the USA, Germany and Russia. Developments of improved iron electrodes have altered the situation, and the iron–nickel oxide system is now being actively considered for EV propulsion and other applications.

The cell in fully charged state can be written as

$$Fe(s)|KOH(aq)|NiO(OH)(s)$$

and has an OCV of about 1.41 V at 25°C (higher when freshly charged due to the presence of higher oxides of nickel). The basic cell reactions are

$$Fe(s) + 2NiO(OH)(s) + 2H_2O(l) \underset{\text{charge}}{\overset{\text{discharge}}{\rightleftharpoons}} Fe(OH)_2(s) + 2Ni(OH)_2 \quad (6.14)$$

It is probable that a range of soluble species such as $Fe(OH)_2^-$ and FeO_2^- are involved, and it is known that $Fe(OH)_3$ or Fe_3O_4 may be formed on deep discharge. The practical energy density of conventional tubular plate cells is 20–30 Wh/kg; with the more recent cells which use press-sintered iron electrodes, values of 40–60 Wh/kg have been reported.

Positive electrodes

The most common configuration is of vertical rows of tubular pocket electrodes held in a nickel-plated steel frame. The tubes are manufactured by spirally winding a perforated plated steel ribbon. The tubes are reinforced by plated steel rings placed at fixed distances along their length, and are packed with alternating layers (> 30 cm) of dried nickel hydroxide powder and nickel flakes ($\approx$ 13%). The latter, which improve the electronic conductance, are prepared from an electrochemically formed multilayer sheet of copper and nickel which is cut and then immersed in sulphuric acid to dissolve out the copper. The tubes are closed and pressure-inserted or welded into the frame to form a unit similar in construction to the $Pb–PbO_2$ tubular system. These electrodes are expensive and difficult to manufacture, but they are very rugged and can readily withstand the stress caused by expansion of the active material. Their life can exceed 7 years, even with heavy cycling duty.

Recently, cells employing thick sintered nickel plates on nickel-plated porous steel substrates have been developed which have greatly improved energy densities. The active material is introduced by electroprecipitation. Electrodes based on nickel fibre supports are also being studied.

Negative electrodes

In pocket plate cells, the active materials are a mixture of finely powdered metallic iron and Fe_3O_4. The preparation of this mixture varies from manufacturer to manufacturer, but generally involves a final process in which controlled air oxidation of iron powder or reduction of Fe_3O_4 with hydrogen is used to form the appropriate composition. Additives such as cadmium, cadmium oxide or graphite are commonly included to improve the capacity retention and electronic conductance. The performance of the electrode is improved by the addition of up to 0.5% of FeS; the mechanism of the sulphide involvement is not well understood. If sulphide is lost by oxidation after prolonged use, small amounts of soluble sulphide may be added to the electrolyte.

Electrolyte

The electrolyte is aqueous KOH with a density of approximately 1.22–1.30 g/cm^3 at 25°C, with 1–2% LiOH addition, as for nickel–cadmium cells.

Cell construction

Iron–nickel oxide cells are always vented. Tubular/pocket plate electrodes are constructed as described above and are generally housed in nickel-plated steel cases. Cells with sintered plate electrodes have smaller inter-electrode spacings. They use synthetic fibre fabrics as separators, and plastic containers.

Performance and applications

The theoretical energy density of the cell is 268 Wh/kg. A typical charge/discharge curve at ambient temperatures at the C/3 rate is shown in Fig. 6.20. A mean discharge voltage of about 1.2 V is realized. Commercial cells normally have capacities in the 250-600 Ah range. Cycle efficiencies are rather low because of the need for extensive overcharge. Low temperature performance is poor compared with that of nickel–cadmium cells, but the most important handicap of this system is its high rate of self-discharge: on 1 month's storage at ambient temperature, 30–50% capacity is lost.

Iron–nickel oxide batteries have been used for many years in railway lighting applications, and for motive power in industrial trucks, tractors and mine locomotives. There is some utilization of the system in emergency lighting and alarm circuits.

Batteries based on the iron–nickel oxide system are now being developed for electric vehicle applications. These use fibre-plaque electrodes, as described above for the nickel–cadmium system, and incorporate electrolyte circulation systems to permit removal of gases evolved during charge

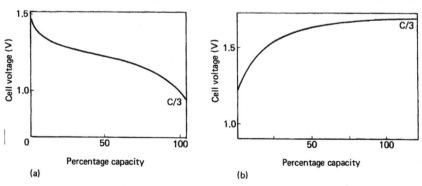

Fig. 6.20 Discharge (a) and charge (b) characteristics of a typical iron–nickel oxide cell at C/3, as a function of percentage capacity

and for thermal management. Advanced iron–nickel oxide batteries have an energy density of over 50 Wh/kg at the C/3 rate and a power density of over 100 W/kg. Electric cars fitted with such batteries have achieved 200 km on a single charge and a cycle life of 1000–2000 has been attained.

Cells with ternary and buffered electrolytes have been studied, as have electrolytes with various additives. The disparity in charging efficiencies between the two electrodes is a significant technical problem still to be solved.

6.6   Zinc–nickel oxide cells

Introduction

Secondary battery systems with zinc electrodes in alkaline solution have been in existence for 100 years, and in the 1930s the Drumm railway battery, an early zinc–nickel oxide power source, was used for railway traction on a regular passenger line in Ireland. However, problems associated with the zinc electrode reduced commercial interest in the system, so that 10 years ago no zinc–nickel oxide battery was being manufactured. Yet many now consider this system to be the most promising short-term solution to the problem of power supply for EV propulsion. The zinc–nickel oxide cell has a number of important advantages, in particular its high operating voltage and energy density, good high rate discharge characteristics and low cost. The theoretical specific capacity of zinc (825 Ah/kg) is almost as high as that of iron and over three times that of lead. The main disadvantage of the system is its relatively poor cycle life and charge retention.

The charged cell, which may be represented as

$$Zn(s)|KOH(aq)|NiO(OH)(s)$$

has an OCV of 1.854 V at 25°C (higher when freshly charged due to the presence of higher oxides of nickel). The cell reactions are, formally,

$$Zn(s) + 2NiO(OH)(s) + 2H_2O(l) \underset{charge}{\overset{discharge}{\rightleftharpoons}} 2Ni(OH)_2(s) + Zn(OH)_2(s) \tag{6.15}$$

However, the discharge product at the zinc electrode is mainly the soluble zincate species $Zn(OH)_4^{2-}$.

Cell construction

Pocket plate design is not suitable for the positive electrode because of the infiltration of soluble zincate and the consequent decrease in positive electrode capacity. Porous matrix positives do not suffer so badly from this problem. Conventional sintered plate positives are relatively heavy and expensive because of the quantity of nickel required as support material. For zinc–nickel oxide EV cells, new techniques have been developed to produce a polymer-bonded matrix using rolling or pressing methods. Graphite is commonly added to improve the electronic conductance, and cobalt to improve capacity retention. Positive electrodes based on nickel fibres have also been studied.

The recharging of zinc electrodes has been widely studied as the development of zinc–air, zinc–silver oxide and zinc–nickel oxide secondary batteries has progressed. The main problems concern the morphology and distribution of the electrodeposited zinc. Under certain conditions, zinc is deposited as dendrites which can penetrate separators and short-circuit the cell. Alternatively, poorly adherent deposits are formed so that the capacity of the electrode is progressively reduced. Further, 'shape changes' may occur as zinc is stripped from the electrode edges during discharge and redeposited near the electrode centre during charge.

To minimize the chance of dendritic growth, it is important to ensure that the zincate ion concentration never falls to very low values. This is accomplished by making cells nickel-limited. Since NiO(OH) electrodes require 20–30% overcharge for full capacity (due to oxygen evolution), it is general practice to make the capacity of the zinc electrode at least twice that of the positive electrode. In sealed cells, oxygen evolved at the positive recombines with zinc at the negative electrode, or is electrochemically reduced so that zincate ion concentration is not depleted even on prolonged overcharge. However, severe overcharging in vented cells can result in overdeposition of zinc: under these circumstances the cell must be short-circuited to restore the zincate concentration.

Investigations are being made into alternative zinc electrode configurations. Contoured electrodes containing PTFE powder have been found to extend cell life, and the use of surface-active agents has been reported to be successful.

Agitation of the electrolyte by a rotating sector or mechanical vibration of the zinc during charge leads to the deposition of more adherent compact layers with a reduction in shape and dendrite problems.

One of the most vital aspects of cell design in their zinc–nickel oxide system is the development of durable separators. Cellulose materials swell in aqueous KOH, causing a tight packing in the cell which helps to preserve the integrity and shape of the zinc electrode. However, oxygen evolved from the positive electrode on charge can cause degradation of cellulose, leading to production of CO_2 and carbonation of the electrolyte. A range of alternative organic and microporous inorganic separators is being studied.

Prototype sealed button cells and larger prismatic cells have been fabricated and studied. As with the nickel–cadmium and nickel–metal hydride systems, an oxygen recombination route is necessary, but the use of membrane separators limits oxygen transport to the negative plate and catalytic recombination devices may be employed. Yuasa is developing a sealed system for powering electric lawn mowers.

Performance and applications

Charge/discharge cycle characteristics for a cell operating at room temperature are shown in Fig. 6.21. The high cell voltage under load (> 1.5 V even at a rate of 2 C) is noteworthy.

Almost all current development of this system is directed towards EV applications where the low cost of the active materials is a significant advantage. Prototype full-scale batteries have been tested by a number of companies in the USA, Europe and Israel. Energy densities of up to 75 Wh/kg (155 Wh/dm^3) can be produced by 300 Ah cells with power densities of

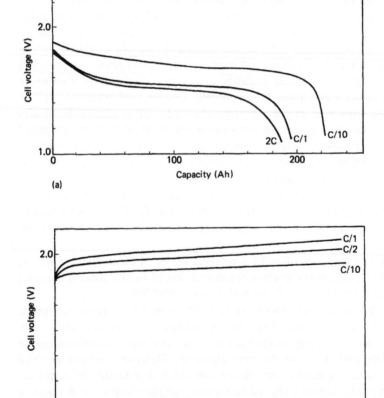

Fig. 6.21 Discharge (a) and charge (b) characteristics of a typical 200 Ah zinc–nickel oxide cell as a function of rate

200 W/kg. Road tests with small passenger vehicles have achieved ranges of 100–230 km on a single charge.

6.7 Zinc–silver oxide and related cells

Introduction

The zinc–silver oxide couple has been known for over 100 years and forms one of the highest energy aqueous cells. The theoretical energy density is 300 Wh/kg (1400 Wh/dm^3) and practical values in the range 40–110 Wh/kg (116–320 Wh/dm^3) have been achieved. Primary cells were considered briefly in Chapter 3.

The first successful secondary zinc–silver oxide alkaline battery system was developed in France by André in the late 1930s. He overcame the problem of migration of silver [as $Ag(OH)_2^-$, $Ag(OH)_3^{2-}$ and similar species] to the zinc electrode by using a cellophane membrane separator. Later developments of this system have emphasized the fundamental role played by separator characteristics in producing a cell of acceptable reliability and cycle life. Apart from high cost and relatively poor cycle life, zinc–silver oxide cell performance deteriorates rapidly below 10°C. However, at higher temperatures the cells can sustain high discharge currents without significant polarization, and their high energy density makes them very suitable for aerospace and certain military applications.

The fully charged cell may be written as

$$Zn(s)|KOH(aq)|AgO(s)|Ag(s)$$

The cell reaction takes place in two stages:

$$2AgO(s) + Zn(s) + H_2O(l) \underset{charge}{\overset{discharge}{\rightleftharpoons}} Ag_2O(s) + Zn(OH)_2(s) \quad (6.16)$$

and

$$Ag_2O(s) + Zn(s) + H_2O(l) \underset{charge}{\overset{discharge}{\rightleftharpoons}} 2Ag(s) + Zn(OH)_2(s) \quad (6.17)$$

with associated emf values of 1.85 and 1.59 V at 25°C, respectively. Both zinc and silver soluble species are also involved in the cell process.

Cell construction

While there are a number of methods used for manufacturing the positive electrodes, the two most important processes are the sintering of silver powders and slurry pasting. The former procedure produces electrodes with superior mechanical properties. The silver mass which is formed by sintering silver powder at temperatures between 400 and 700°C is supported on silver or silver-plated copper grids. This method allows a continuous manufacturing process and a final assembly with the positive electrode in a fully discharged condition – which greatly improves the storage characteristics of the system. Organic resins and pore-forming materials may be added to the silver and continuous sheets formed by a rolling process. This method is used when thin sintered plates are required. Non-sintering methods are based on electroforming plates which have been prepared by spreading a silver oxide/water slurry onto suitable support grids. In some processes, thermal decomposition at 400°C precedes electroforming. Again, this method of manufacture may be performed on a continuous basis.

Many different fabrication procedures are used for the zinc electrode. One group of methods starts with zinc oxide and various additives and follows with electrochemical reduction in dilute aqueous KOH. In another, zinc is electrodeposited from a cyanide bath. Others use mixtures of zinc, zinc oxide and organic binding agents. Generally the electrodes are formed on open-mesh grids. In all cases the aim is to produce electrodes of high porosity and controlled thickness. Additives include surface active agents to minimize dendritic growth and mercuric ions to increase the hydrogen overvoltage of the zinc and so reduce corrosion.

The separator assembly is the most critical component of zinc–silver oxide secondary cells. In addition to its normal function of preventing contact and short circuit between electrodes of different polarity, a separator in this system must also:

- prevent silver migration to the negative electrode;

- control zincate migration;

- have the swelling properties required to establish favourable zinc morphology and distribution on charge;

- preserve the integrity of the zinc electrode.

In addition, the separator must have a low electrical resistance, good thermal and chemical stability and must be light in order to retain the high energy density characteristics of the cell. Practical separators have a composite multilayer configuration. A 'silver-stopping' layer of cellophane or non-woven synthetic polyamide is located next to the positive electrode which reduces soluble silver species back to the metal. A potassium titanate paper layer may be placed next to the zinc electrode, and a number of cellophane layers which swell in aqueous KOH make up the middle section. In most cells the separators are fabricated as envelopes or 'sacks' which completely enclose the zinc electrodes.

Commercial cells are generally rectangular (prismatic) in shape and the case is usually plastic, with a rugged construction to withstand the mechanical stress to which this system is often subjected. The cells are sealed, but

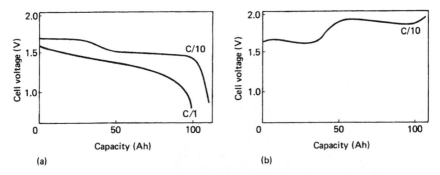

Fig. 6.22 Discharge (a) and charge (b) characteristics of a typical 100 Ah zinc–silver oxide cell

normally have a safety filler valve assembly. (Small quantities of oxygen evolved at the silver electrode during recharge may recombine with excess zinc on the negative electrode; however, the membrane separator assembly slows down oxygen diffusion.) The quantity of free electrolyte is very small: the majority is contained within the electrode pores or separator.

Freshly assembled cells are subjected to a series of formation cycles in order to activate the system. Cells are often sold in a charged but dry state, in which case the formation process is performed before the final cell assembly. Dry charged batteries can be stored indefinitely. Cells containing electrolyte should be stored in the completely discharged state since dissolution of silver oxide is then avoided.

Performance and application

The energy density of practical zinc–silver oxide cells is some five to six times higher than that of their nickel–cadmium equivalents. In Fig. 6.22(a), typical low and high rate discharge curves are shown. At low rates two plateaus appear, at about 1.7 and 1.5 V, corresponding to the reactions given in eqs (6.16) and (6.17), respectively. The change from Ag(II) to Ag(I) controlled processes is dependent on the discharge rate and the upper plateau disappears completely at high rates. Reduction in cell capacity at high rates of discharge is not very significant. However, sustained high rates can result in temperatures which may damage organic separator materials. In Fig. 6.22(b) a typical recharge curve at C/10 is shown (much faster charging is possible). The cell can be considered to be fully charged when its voltage reaches 2.0 V – excessive overcharge should be avoided because of the increased probability of zinc dendrite formation and possible damage to the separator by reaction with oxygen. The main drawbacks of the system are its high cost combined with a poor working and cycle life.

Zinc–silver oxide secondary cells with capacities of 0.5–100 Ah are manufactured for use in space satellites, military aircraft, submarines and

Fig. 6.23 Battery for 'Ranger' lunar photography spacecraft, comprising 14 sealed 45 Ah zinc–silver oxide cells, which impacted on the moon in 1965. (By permission of Jet Propulsion Laboratory.)

for supplying power to portable military equipment. In space applications, the batteries are used to augment the power from solar cells during periods of high demand, e.g. during radio transmission or when the sun is eclipsed. At other times the batteries are trickle or float charged by the solar cells. The battery for the 'Ranger' lunar photography spacecraft, which consisted of 14 sealed 45 Ah cells is shown in Fig. 6.23. This spacecraft impacted on the moon in 1965. In military aircraft the use of zinc–silver oxide batteries as emergency power supplies or to provide additional power for certain manoeuvres requiring rapid actuation of flight control surfaces, may be justified because of their high energy density. 'One-shot' reserve zinc–silver oxide batteries are also used in military aircraft. A few large submarine batteries have also been built using the zinc–silver oxide system as an alternative to lead–acid cells. Heat dissipation problems require that particular attention be given to the way in which the cells are stacked. Submarine batteries of this type are usually stored under oil.

Cadmium–silver oxide cells

Replacing zinc with cadmium reduces the OCV by approximately 0.4 V, but increases the cycle life of the system considerably. This cell is very similar to the nickel–cadmium system, but has an energy density higher by about a third. The cost of the system has restricted its application to small button cells.

Iron–silver oxide cells

The overall cell reaction is exactly similar to the zinc–silver oxide cell, with the final discharge products being silver and $Fe(OH)_2$. The emf of the cell is only 1.34 V, but the cell has a superior cycle life and better reliability. Batteries up to 10 kWh have been built by Westinghouse for telecommunications and submersible vehicles.

Metal hydride–silver oxide cells

The overall cell reaction may be written as

$$2M(s) + Ag(s) + H_2O(l) \overset{\text{charge}}{\underset{\text{discharge}}{\rightleftharpoons}} 2MH(s) + AgO(s) \qquad (6.18)$$

although again the silver electrode undergoes two sequential reactions. This cell system combines the high energy density of the silver electrode with the long cycle life of the metal hydride and thus potential improvement in performance over nickel–metal hydride and zinc–silver oxide cells. Development work is being led by Eagle-Picher with a view to producing batteries for aerospace and military applications. Consideration is also being given to using this system in high value commercial applications.

7 Rechargeable lithium cells

Bruno Scrosati

7.1 Introduction

Because of the limited energy density of the active electrode components in aqueous cells based on the lead–acid or nickel–cadmium systems, conventional batteries do not meet the requirements of modern consumer electronic devices, nor those of electric traction. More recent developments such as the nickel–metal hydride system described in Chapter 6 provide improved performance, but in order to achieve significant increases in energy density it has proved necessary to develop advanced batteries based on alkali metal anodes. High capacity systems designed primarily for electric vehicle applications and which operate at high temperatures are considered in Chapter 8. These systems are not suitable for consumer electronics. In this chapter, we describe the development and commercialization of ambient temperature, high energy secondary batteries based on lithium. These are currently available in the consumer market in the form of standard button and cylindrical cells, but there is now a major nationwide effort to produce improved systems, and in particular to develop reliable high capacity systems for EV traction. Fig. 7.1 shows the relative position of rechargeable lithium batteries in 1997 in comparison with aqueous batteries in terms of gravimetric and volumetric energy density, together with projected values for future systems.

Initial development of ambient secondary lithium batteries was based on the primary lithium systems described in Chapter 4, consisting of a lithium metal negative, a non-aqueous lithium ion conducting electrolyte and a positive electrode material which could undergo a *reversible* electrochemical reaction with lithium ions:

$$x\text{Li}^+ + \text{A}_z\text{B}_y(\text{s}) + xe \rightleftharpoons \text{Li}_x\text{A}_z\text{B}_y(\text{s}) \tag{7.1}$$

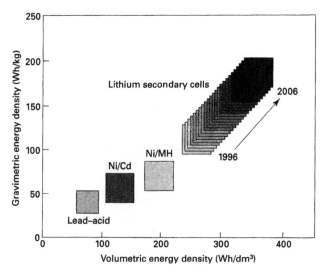

Fig. 7.1 Comparison of gravimetric and volumetric energy density of lithium secondary cells with aqueous electrolyte-based systems

Suitable positive electrode materials A_zB_y, such as TiS_2 or V_6O_{13}, generally have rigid lattices with:

- layer, tunnel or open structures which permit insertion (or extraction) of lithium ions without major structural modification;

- the ability to accept (or eject) compensating electrons into (or from) their electronic bands.

These materials are known as insertion or intercalation hosts. The overall electrochemical process of a lithium battery is illustrated schematically in Fig. 7.2. During discharge it involves the dissolution of lithium ions at the anode, their migration across the electrolyte and their insertion within the crystal structure of the host compound, while the compensating electrons travel in the external circuit to be injected into the electronic band structure of the same host. The charging process is the reverse and the cell reaction may be written as:

$$x\text{Li(s)} + A_zB_y\text{(s)} \underset{\text{charge}}{\overset{\text{discharge}}{\rightleftharpoons}} \text{Li}_xA_zB_y\text{(s)} \qquad (7.2)$$

As discussed below, there are problems with morphological changes and passivation reactions at lithium metal negative electrodes in secondary cells, which reduce cycle life and the practical energy density of the system, and may in some circumstances introduce safety hazards. A more recent development involves the replacement of the lithium metal anode by another insertion compound, say C_nD_m. In this cell, the electrochemical process at the negative side, rather than lithium plating and

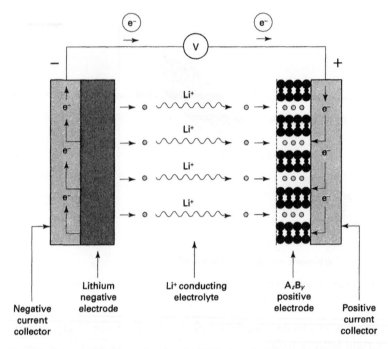

Fig. 7.2 Schematic diagram showing the discharge of a lithium–A_zB_y intercalation positive cell. (By permission of *Chim & Ind.*: B. Scrosati, 1995, **77**, 285.)

stripping as in a conventional system, is the uptake of lithium ions during charge and their release during discharge (Fig. 7.3). Therefore, during discharge the negative $Li_xC_nD_m$ electrode acts as a lithium ion source while a positive A_zB_y electrode acts as a lithium ion sink and the total electrochemical process of the $Li_xC_nD_m/A_zB_y$ cell involves the cyclic transfer of x equivalents of lithium ions between the two insertion electrodes:

$$Li_xC_nD_m(s) + A_zB_y(s) \underset{charge}{\overset{discharge}{\rightleftharpoons}} C_nD_m(s) + Li_xA_zB_y(s) \quad (7.3)$$

These systems have been termed 'rocking chair batteries', or more recently with a series of alternative names, such as 'swing' or now more generally 'lithium ion' batteries. In Fig. 7.4 the approximate voltage ranges for reversible operation of a number of lithium insertion hosts are given, from which it is seen that the cell voltage of a lithium ion cell will be only about half that of a cell with a lithium metal negative. Such a reduction in the energy is not acceptable, and most current development of lithium ion cells uses a carbon-based negative, for which there is a reversible electrochemical reaction given by

$$xLi^+ + 6C(s) + xe \rightleftharpoons Li_xC_6(s) \quad (7.4)$$

As $x \rightarrow 1$, the voltage of the Li_xC_6 approaches 0.02 V with respect to

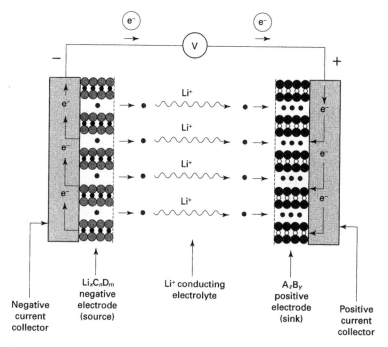

Fig. 7.3 Schematic diagram showing the discharge of a $Li_xC_nD_m$–A_zB_y cell in which both positive and negative electrodes are based on intercalation hosts. (By permission of *Chim. & Ind.*: 1995, **77**, 285.)

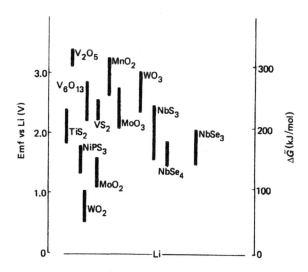

Fig. 7.4 Voltage ranges for reversible operation of some insertion hosts, after F.A. Trumbore. (By permission of *Pure and Applied Chemistry*: 1979, **52**, 119.)

lithium, so that the cell voltage is not significantly different from that of lithium metal-based cells. The overall cell reaction for a lithium ion system with a graphitic negative is then

$$Li_xC_6(s) + A_zB_y(s) \underset{\text{charge}}{\overset{\text{discharge}}{\rightleftharpoons}} 6C(s) + Li_xA_zB_y(s) \qquad (7.5)$$

7.2 Negative electrodes

Lithium metal

The properties of lithium metal were described in Chapter 4, where particular note was made of its high specific capacity and electrode potential. However, because of its highly electropositive nature, it is thermodynamically unstable in contact with a wide variety of reducible materials. In particular, lithium reacts with components of most electrolytes to form a passivating layer. Film formation of this type ensures long shelf life for primary lithium cells, but causes severe problems when the electrode is cycled in a secondary cell.

It is found that lithium can be plated with virtually 100% efficiency in a range of organic systems; however, the plated lithium cannot be stripped quantitatively, especially if the cell has been allowed to stand for a period between plating and stripping. An explanation for this behaviour, advanced by Brummer of EIC Corporation, is illustrated in Fig. 7.5. The lithium is considered to be electrodeposited in granular form and the newly created surfaces react rapidly with components of the electrolyte; this continues

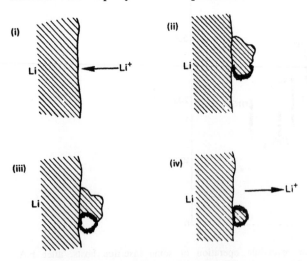

Fig. 7.5 Mechanism for the isolation of plated lithium by an insulating film, as originally suggested by S.B. Brummer of EIC Corporation

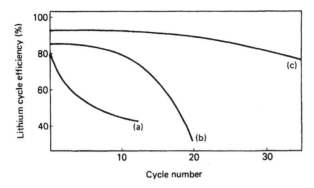

Fig. 7.6 Effect of electrolyte composition on lithium cyclability, after V.R. Koch *et al.* of EIC Corporation. (a) $LiClO_4$–PC; (b) $LiAsF_6$–PC; (c) $LiAsF_6$–2Me-THF

once the charging (plating) current has been switched off. Some lithium grains become partially undercut and others are completely isolated from the underlying lithium metal by an insulating film. Discharge (stripping) efficiency is therefore less than 100% and the residual isolated lithium grains affect the morphology of any subsequent replating. After a few cycles, the capacity or Ah efficiency of the cycle may fall to almost zero.

It was found that by altering the constituents of the electrolyte, the nature of the passivating film could be modified to such an extent that the cycling behaviour was greatly improved. Attempts were therefore made to find an optimized electrolyte system which would result in the formation of 'desirable' films, i.e. films which would be impermeable to solvent and stabilize the metal, but which would remain conductive to lithium ions. One of the best electrolytes to date contains $LiAsF_6$ dissolved in carefully purified 2-methyltetrahydrofuran (2-Me-THF). This solvent was selected because of the relatively low polarity of the C–O ether bond, and because of the predicted effect of the methyl group on slowing the formation of ring-opened products by the lithium. The effect of changing the electrolyte constituents is shown in Fig. 7.6. However, the exact reasons for the excellent behaviour of the $LiAsF_6$/2-Me-THF system are not fully understood. An experimental cell based on this electrolyte sustained over 100 deep cycles, but some loss in stripping efficiency is still evident when cells are allowed to stand for long periods at open circuit.

A phenomenon known as thermal runaway can occur when a large new lithium surface is produced on charge which reacts exothermically with components of the electrolyte. In some cases the temperature is raised sufficiently to melt some of the lithium, which then in turn reacts with more electrolyte, causing a further rise in temperature and eventually the device goes on fire.

Carbon-based negative plates

As noted above, lithium ions may be reversibly inserted into a variety of carbonaceous materials, according to eq. (7.4). Carbons which have been

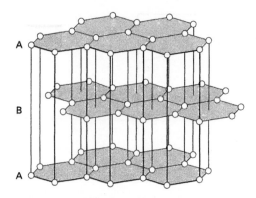

Fig. 7.7 Structure of graphite

studied include natural and synthetic graphites, petroleum coke, carbon fibres and mesocarbons, which differ, depending on temperature and method of preparation, in degree of crystallization and stacking order. The amount of lithium which can be reversibly inserted into the host carbon structures, i.e. the value of x in Li_xC_6, depends on a number of factors. The structure of graphite, shown schematically in Fig. 7.7, is formed by planar layers of carbon atoms stacked in an ABABAB . . . sequence. Natural graphites present various amounts of stacking disorder, while some natural and synthetic carbons may be characterized as having very disordered structures with random stacking.

Lithium intercalation occurs between the carbon planes. At ambient temperatures, pure graphite intercalates up to one lithium per six carbon atoms to form the compound LiC_6. LiC_6, which can also be prepared chemically, has a golden colour and is known as a stage 1 graphite insertion compound (GIC), where the stage number corresponds to the number of graphite layers which separate two successive intercalated planes. The maximum attainable capacity of the LiC_6 electrode is 0.372 Ah/g. The fabrication of practical carbon electrodes requires the addition of a binder, such as ethylenepropylenediene monomer (EPDM), polyvinylidene fluoride (PVDF) or polytetrafluoroethylene (PTFE). The final electrode is generally formed from a slurry of carbon, binder and a suitable solvent by a casting or pressing procedure.

The electrochemical response of a given carbon electrode may be evaluated by following the lithium intercalation process in a cell using a lithium metal counter electrode, i.e. by monitoring the potential of the carbon electrode during intercalation or de-intercalation. Fig. 7.8 illustrates a typical voltage–capacity curve for a graphite electrode, determined during the first intercalation–deintercalation cycle. The voltage decreases in a series of readily distinguishable plateaus which correspond to the progressive formation of staged phases, ranging from LiC_{32} to LiC_{12}, to finally reach LiC_6 at a voltage approaching 0.02 V with respect to lithium.

The charge–discharge profile of amorphous carbon (coke) electrodes is

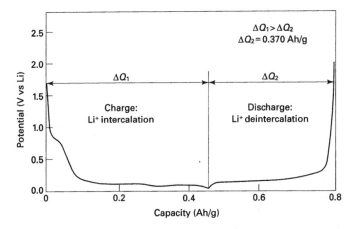

Fig. 7.8 Potential–capacity curve for a graphite electrode, determined during the first intercalation/de-intercalation cycle

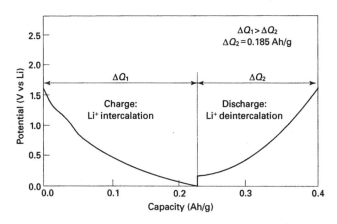

Fig. 7.9 Potential–capacity curve for a coke electrode, determined during the first intercalation/de-intercalation cycle

somewhat different from that observed in graphite electrodes. As shown in Fig. 7.9, the cycling profile of these structurally disordered carbon electrodes shows no evidence of staging plateaus, but rather continuous charge–discharge curves sloping between 1.2 and 0.2 V and which cannot be extended beyond a composition of approximately $Li_{0.5}C_6$. This depresses the capacity of the coke electrode to an average of 0.186 Ah/g, with differences which depend on the thermal history of the sample.

One can see from both Figs 7.8 and 7.9 that the charge consumed in this first charging process, ΔQ_1, exceeds the expected maximum theoretical capacities (i.e. 0.372 Ah/g and 0.186 mAh/g for graphite and coke, respectively) for a composition of LiC_6 or $Li_{0.5}C_6$. This formal excess in capacity, which is typical for all types of carbon electrodes, is caused by side reactions involving the decomposition of the electrolyte which induce the formation of a passivating film on the electrode surface. This layer is found to be electronically insulating but ionically conducting. It thus prevents further electrolyte decomposition but allows ionic transfer with the solution. In the discharge (lithium release) immediately following the first charge, as well as in all the subsequent charge–discharge cycles, the capacity of the electrode, ΔQ_2, approaches its expected value and remains essentially stable with time and cycle number. The composition and characteristics of the passivation film depend on the electrolyte components. Its protective action against further electrolyte decomposition allows operation at voltages ranging between 0.25 and 0.0 V versus lithium. Unlike the situation with metallic lithium discussed above, there is no further film formation on subsequent cycles.

Obviously, the replacement of lithium metal by Li_xC_6 induces penalties in energy density due to losses in specific capacity: 0.186 Ah/g for cokes ($x = 0.5$) and 0.372 Ah/g for graphite ($x = 1$) in comparison with 3.86 Ah/g for Li metal. One has to recall, however, that because of its relatively poor ability to cycle, an excess of lithium, generally four times the stoichiometric amount, is incorporated in practical lithium metal-based secondary cells in order to assure acceptable cycle life, and this somewhat offsets the difference. In addition, there are indications that by a proper selection of the type of carbon and/or by the incorporation of additives in the carbon structure, the capacities of the Li_xC_6 electrodes may be extended beyond the 0.372 Ah/g value by exploiting intercalation mechanisms other than staging. Boron-substituted metastable $LiC_{3.8}$ and $LiC_{2.2}$ high capacity phases have been reported which, however, revert in time to the common LiC_6 phase. Other approaches involve the synthesis of highly disordered graphite and of non-graphitic carbons where an excess of lithium can be inserted to reach higher capacities. A variety of carbon-based materials have been prepared either by low temperature treatments of resins or by chemical vapour deposition. The micro- and nanoporosity of these materials can be controlled by oxidation, leading to electrodes capable of intercalating lithium in excess of 0.600 Ah/g with a mechanism involving adsorption of lithium on the internal surface of the nanopores. The size of the latter can be controlled in such a way as to prevent solvent penetration and hence large irreversible capacity losses. Materials prepared by chemical vapour deposition are often doped with metals capable of alloying lithium. For instance, it has been reported that nanodispersed silicon atoms can reversibly bond up to about 1.5 lithium ions. The carbonaceous matrix, characterized by a large degree of turbostratic disorder, provides diffusion pathways to the nanodispersed silicon atoms while retaining its own ability to intercalate lithium. Although the stability of these new materials on cycling has not yet been fully established, it is clear that electrodes with capacities exceeding 0.370 Ah/g may be available in the near future. Ideal carbon electrodes for lithium ion batteries should have minimum initial irreversible capacity (less than 10% of total carbon capacity), relatively low surface area (less than 5 m^2/g), high packing density (more than 2 g/cm^3) and a high practical capacity for cycling.

In addition to carbon-based systems, other intercalation compounds are also currently being proposed as alternative lithium ion cell negative plates. Examples include Li_xTiS_2, Li_xTiO_2, $Li_{3+x}Ti_5O_{12}$ and, more recently, a family of Li_xSnO_y compounds. However, the applicability of these materials in practical batteries has not yet been established, and coke and graphite are still the materials used in all commercial lithium ion cells.

7.3 Positive electrodes

Insertion hosts which have been developed as positive electrode materials may be divided into two categories. The first group, sometimes known as 'low voltage cathodes', have been used principally as the active material in the positive electrodes of lithium metal-based cells. The average voltage of this group is in the range 2.2–3.2 V with respect to lithium. The second group form the 'high voltage cathodes' with voltages around 4 V, and are used in conjunction with carbon-based negatives in lithium ion cells. Note, however, that there is no difference in principle between these two groups of electrode materials.

'Low voltage' positives

Vanadium oxides

The vanadium oxide V_6O_{13} has a structure which can be described as consisting of distorted VO_6 octahedra joined by extensive edge sharing into single and double chains which are linked (by additional edge sharing) into single and double sheets. The sheets are joined together by corners to form a three-dimensional framework of channels through which lithium ions can diffuse. V_6O_{13} can accommodate eight lithium ions per mole according to the process

$$8Li(s) + V_6O_{13}(s) \rightleftharpoons Li_8V_6O_{13}(s) \qquad (7.6)$$

Lithium batteries based on this reaction have an open circuit voltage of around 3.2 V. The voltage varies with composition, with the typical trend illustrated in Fig. 7.10. The Li^+ intercalation process induces phase modification and, although the process is reversible over the entire composition range, long cyclability of the cell is readily achieved only when x is limited to 4, i.e. to formation of $Li_4V_6O_{13}$. The theoretical energy density of the Li/ V_6O_{13} couple can be calculated from eq. (7.6) assuming an average operating voltage of 2.4 V. The value, based on the active mass of the reactants, is of the order of 860 Wh/kg.

Other forms of vanadium oxide have been used as cathodes in rechargeable lithium batteries. A common example is V_2O_5 which displays a structure consisting of distorted VO_6 octahedra joined by edge sharing

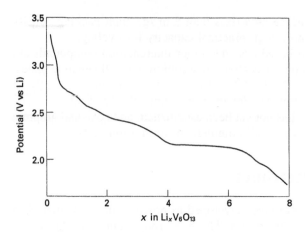

Fig. 7.10 Potential–composition curve of a $Li_xV_6O_{13}$ electrode

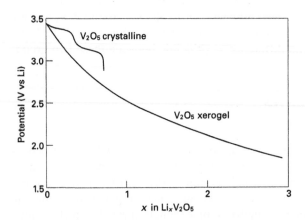

Fig. 7.11 Potential–composition curve of $Li_xV_2O_5$ electrodes

into single zig-zag chains linked together by corner sharing to form single sheets. The sheets are joined by means of additional corner sharing into a layered structure, with layers formed by a sequence of VO_5 pyramids and connected by oxygen bridges. The lithium intercalation process is

$$xLi(s) + V_2O_5(s) \rightleftharpoons Li_xV_2O_5(s) \qquad (7.7)$$

where x depends on the morphology of the vanadium oxide. As shown in Fig. 7.11, the value of x varies from 0.4 in the case of crystalline electrodes to more than 3 for xerogel film electrodes. Accordingly, the theoretical energy density ranges from 460 to 1130 Wh/kg.

Another useful member of the vanadium oxide group is the LiV_3O_8 vanadium bronze. Its basic structure is formed by octahedral and trigonal bipyramids arranged to form puckered layers between which the Li^+ ions

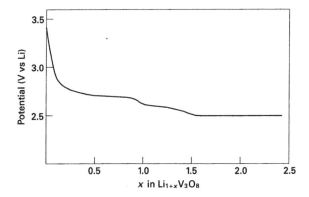

Fig. 7.12 Potential–composition curve of a $Li_{1+x}V_3O_8$ electrode

are situated. The unit cell comprises six empty tetrahedral sites which may easily accommodate up to three lithium ions per mole:

$$3Li(s) + LiV_3O_8(s) \rightleftharpoons Li_{1+3}V_3O_8(s) \qquad (7.8)$$

The OCV of Li/LiV_3O_8 cells is approximately 3.3 V at room temperature and the voltage/composition curve is shown in Fig. 7.12. Single-phase behaviour, typical of solid solution formation, is only visible up to an x value of about 1.5. At a higher Li^+ content, a structural rearrangement of the vanadium bronze takes place. Repulsion between the lithium ions causes ordering into specific sites so that a new phase appears which is in equilibrium with that corresponding to the upper limit of the solid solution and thus the potential remains constant, as is typical of a two-phase region (see Chapter 2). It has been shown that these changes in structure are highly reversible and thus that the vanadium bronze electrode can be repeatedly cycled over the full $0 < x < 3$ composition range. Assuming an average operating voltage of 2.8 V, the theoretical energy density for the Li/LiV_3O_8 couple is 740 Wh/kg.

Titanium disulphide

Historically the most commonly used lithium intercalation compound was titanium disulphide, TiS_2. This compound has a layered structure of covalently bound S–Ti–S stacks held together by weak van der Waals forces. Each stack is formed by a layer of titanium atoms between two layers of sulphur atoms in a hexagonally close packed arrangement. Lithium ions can be readily intercalated between the stacks, and if the intercalation level x is maintained below unity, the process induces only a modest and reversible expansion along the c axis (Fig. 7.13). The electrochemical reaction of the Li/TiS_2 couple

$$Li(s) + TiS_2(s) \rightleftharpoons LiTiS_2(s) \qquad (7.9)$$

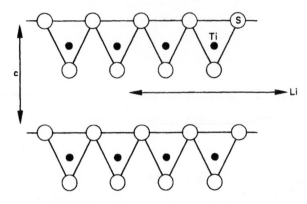

Fig. 7.13 Schematic structure of TiS$_2$ layers

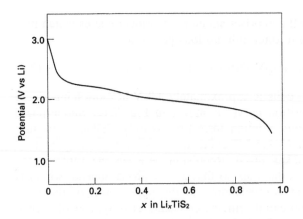

Fig. 7.14 Potential–composition curve of a Li$_x$TiS$_2$ electrode

has an OCV of 2.6 V at room temperature. The voltage decreases with concentration with a continuous trend (Fig. 7.14), typical of a single-phase reaction. Assuming an operating voltage of 2.2 V, one obtains a theoretical energy density of 480 Wh/kg.

Molybdenum disulphide

Molybdenum disulphide is another layered intercalation host, similar to titanium disulphide. This material occurs naturally and formed the basis of the positive electrode for the first high production cylindrical AA-sized cell, manufactured by Moli Energy Ltd in Canada in the 1980s. Cycle life of 100–300 was achieved in practical cells with average discharge voltages of 1.8 V for low rates, giving a theoretical density of approximately 300 Wh/kg.

Manganese dioxide

Manganese dioxide is a cheap, readily available material which is used worldwide as a solid cathode for primary liquid cells. During recent years, remarkable progress has been made in the development of new synthetic manganese dioxide phases. Of particular interest are the $LiMn_2O_4$ compounds which are available in two forms – a 'high voltage' (about 4 V with respect to Li) $Li_{(1-x)}Mn_2O_4$ form which is capable of releasing lithium ions and which will be discussed below; and a 'low voltage' (about 3 V with respect to Li) $Li_{(1+x)}Mn_2O_4$ form which is capable of accepting lithium ions. This compound has a cubic spinel structure in which the Mn_2O_4 framework is constructed of face sharing octahedra and tetrahedra which provide a conducting pathway for the lithium ions. The structure remains intact for lithium insertion and extraction over the $-1 < x < 1$ composition range. Fig. 7.15 shows the voltage–composition curve for the $Li_{1+x}Mn_2O_4$ low voltage electrode. The lithium insertion induces a Jahn–Teller distortion which reduces the crystal symmetry from cubic in $LiMn_2O_4$ to tetragonal in $Li_{1+x}Mn_2O_4$, thus resulting in a two-phase system which produces a desirable flat voltage of 2.9 V. Using this as the average working voltage value and assuming $x = 1$, the theoretical energy density, based on the electrochemical reaction

$$(1-x)Li(s) + Li_{1+x}Mn_2O_4(s) \rightleftharpoons Li_2Mn_2O_4(s) \qquad (7.10)$$

is 440 Wh/kg.

Electronically conducting polymers

It is perhaps useful to mention that the use of electronically conducting polymers, such as poly(acetylene), $(CH)_x$, poly(pyrrole), $(C_4H_5N)_x$, and poly(aniline), $(C_6H_5NH_2)_x$, has been proposed for positives for lithium batteries. The electrochemical process of these 'lithium–polymer' positives is somewhat similar to an intercalation reaction. On charging, the polymer (P) is oxidized by acquiring a positive charge to form a polaron, and this is

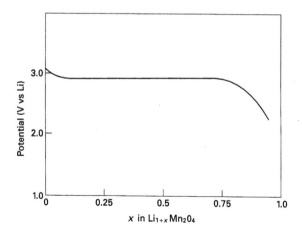

Fig. 7.15 Potential–composition curve of a 'low voltage' $Li_{1+x}Mn_2O_4$ electrode

compensated by the insertion of a negative ion from the electrolyte. The electrochemical process of a cell with a lithium negative may be written as

$$P(s) + xy LiX \text{ (electrolyte)} \underset{\text{discharge}}{\overset{\text{charge}}{\rightleftharpoons}} [P^{y+}](yX^-)_x(s) + xy Li(s) \quad (7.11)$$

where P is the polymer (e.g. poly(pyrrole)) cathode and LiX is the lithium salt.

The charging process is sometimes called doping, and the amount of charge involved, namely y in eq. (711), is the doping level of the polymer P. Clearly, the higher the doping level, the higher is the capacity of cell. Reasonable capacities and good cyclability have been reported for some prototypes. However, interest in these batteries has fallen due to their comparatively low energy density and due to poor charge retention as a result of self-discharge.

Recently, new types of organic positives, formed by a mixture of dimer-captan with polyaniline, have been reported. These composite materials appear to have an energy density higher than that of most inorganic intercalation oxides and good cyclability. However, use of these materials in practical rechargeable lithium batteries has still to be confirmed.

The theoretical energy densities of couples based on the most important 'low voltage' positive materials together with lithium are given in Table 7.1. It is generally assumed that the energy density of practical batteries is only 20–25% of the theoretical values based solely on the masses of the active electrode components. Applying this empirical rule, it is seen that energy densities of 80–100 Wh/kg can be predicted for lithium batteries, as

Table 7.1 Theoretical energy density of Li/A_zB_y electrochemical couples based on the xLi + $A_zB_y \rightarrow Li_xA_zB_y$

A_zB_y	Composition range	Average voltage (V)	Energy density (Wh/kg)
V_6O_{13}	$0 < x < 8$	2.4	860
V_6O_{13}	$0 < x < 3.6$	2.4	415
V_2O_5	$0 < x < 1$	3.2	460
V_2O_5 (xerogel)	$0 < x < 3$	2.5	1130
LiV_3O_8	$0 < x < 3$	2.8	740
TiS_2	$0 < x < 1$	2.2	480
MoS_2	$0 < x < 1$	1.8	300
$LiMn_2O_4$	$0 < x < 1$	2.9	440
$(C_4H_5N)_x$	$y < 0.25$	3.0	240

Table 7.2 Characteristics and performance details of practical Li/A$_z$B$_y$ rechargeable battery prototypes

A$_z$B$_y$	Size	Capacity (Ah)	Manufacturer	Energy density		Cycle life
				(Wh/kg)	(Wh/dm^3)	
MoS$_2$	AA	0.6	Moli Energy Ltd	50	135	400
MnO$_2$	AA	0.75	Moli Energy Ltd	120	265	400
TiS$_2$	AA	0.9	EIC Lab. Inc. W.R. Grace	95	235	250
V$_2$O$_5$*	Coin	0.03	Panasonic Ind. Co.	30	100	1000
LiMn$_3$O$_6$	AA	0.75	Tadiran Ltd	130	280	300

* Li–Al alloy negative plate.

is confirmed by experimental results (Table 7.2). Details of these practical battery systems are considered in Section 7.5.

'High voltage' positives

The most common 'high voltage' materials for positive plates are the layered lithium metal oxides LiCoO$_2$ or LiNiO$_2$, and the three-dimensional, spinel-type LiMn$_2$O$_4$. The layered oxides have a rock salt structure where lithium and transition metal cations occupy alternate layers of octahedral sites in a distorted cubic close-packed oxygen ion lattice (*see* Fig. 7.16), and the framework provides a two-dimensional interstitial space which allows for easy extraction or insertion of lithium ions. LiCoO$_2$ can be prepared by heating a pelletized mixture of lithium hydroxide and cobalt carbonate in air at 850°C, while LiNiO$_2$ is obtained by annealing an intimate mixture of Li$_2$O and NiO$_2$ at 850°C. Regardless of the method of preparation, the LiMO$_2$ compounds may exhibit non-stoichiometry, generally due to an excess of M. Since the excess M occupies structural sites otherwise available for lithium ions, the excess stoichiometry affects the specific capacity and the electrochemical response of the LiMO$_2$ electrodes. Control of the synthesis conditions is therefore critical in assuring optimum performance. Practical electrode configurations involve mixtures of the LiMO$_2$ powder, carbon and binder, formed into a layer and coating on a metallic (often aluminium) foil or mesh substrate.

Among the LiMO$_2$ compounds, LiCoO$_2$ has attracted particular attention because of its high voltage, 4.5 V with respect to Li, a value which is consistent with the high oxidizing power of the Co^{4+}/Co^{3+} couple. Although at the time of the initial development of lithium ion cells, LiCoO$_2$ was the most popular electrode, it has a number of significant drawbacks, including high cost (cobalt is twice the price of nickel and 10 times that of

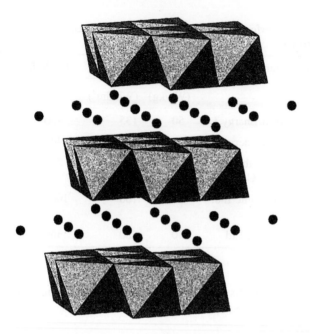

Fig. 7.16 Schematic structure of layered lithium metal oxides. (By permission of Dr A.R. Armstrong, University of St Andrews.)

manganese), a limited range of cycling lithium content (between approximately $Li_{0.45}CoO_2$ and $LiCoO_2$), and the possibility of cointercalation of solvent between the layers. The toxicity of cobalt compounds is a particularly serious problem. The characterization of lithium metal oxide positives is carried out in a similar way to that used for the carbon negative electrodes, namely by monitoring the voltage during lithium intercalation/de-intercalation in cells using lithium metal as counter electrode. A typical voltage composition curve for the $Li_{(1-x)}CoO_2$ electrode is illustrated in Fig. 7.17. It is generally assumed that the de-intercalation of lithium induces phase transitions which are accompanied by lattice distortion. The occurrence of these phase transitions is revealed by the plateau regions in the composition curve of Fig. 7.17. Initial studies on the reversibility of Li_xCoO_2 showed that in the composition range $0.5 < x < 1$, there are three reversible phase transitions. There is also a fourth, irreversible transformation at $x < 0.5$ where the layered rock-salt structure of Li_xCoO_2 is destroyed. This limits the practical cyclability of $LiCoO_2$ to $x = 0.5$, with an associated charge–discharge voltage range of 4.5 to 3.0 V with respect to Li and a maximum practical specific capacity of about 180 mAh/g.

$LiNiO_2$ has a structure consisting of cubic close-packed O^{2-} ions with Ni^{3+} ions occupying alternate layers of octahedral sites between adjacent close-packed oxide layers, while Li^+ ions occupy the other set of octahedral sites between the oxide layers. The variation of the voltage of the $Li_{(1-x)}NiO_2$ electrode is shown in Fig. 7.17 as a function of lithium

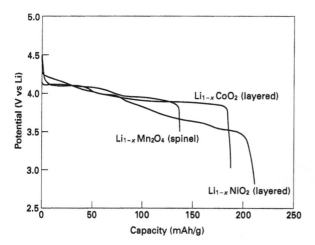

Fig. 7.17 Potential–capacity curves for layered $Li_{1-x}CoO_2$, layered $Li_{(1-x)}NiO_2$ and $Li_{(1-x)}Mn_2O_4$ spinel. (By permission of *J. Power Sources*: S. Megahed and B. Scrosati, 1994, **77**, 79.)

content. Short range ordering of Li^+ ions and electrons has been suggested to account for the plateaus observed in the voltage–composition curve. A maximum capacity of 220 mAh/g can be obtained for the $LiNiO_2$ cathode. $LiNiO_2$ has a higher capacity than $LiCoO_2$ for lithium cycling. Solid solutions such as $Li(Ni_{1-x}Co_x)O_2$ offer a compromise between stability, cost and capacity, and have occasionally been used.

The high voltage form of the spinel $LiMn_2O_4$ is currently regarded as the most attractive of the series. It is less expensive and environmentally much more benign than its rivals. $LiMn_2O_4$ adopts a spinel structure consisting of cubic close-packed oxide ions with the manganese ions in one-half of the octahedral sites and lithium ions in one-eighth of the tetrahedral sites (Fig. 7.18). The Mn_2O_4 framework occupies a three-dimensional region via face sharing octahedra and tetrahedra which provide conducting pathways for the insertion and the extraction of the lithium ions. The potential/composition curve for the $Li_{(1-x)}Mn_2O_4$ electrode is shown in Fig. 7.17. The removal of lithium can be extended to $x = 1$. The curve displays two plateaus (each associated with a two-phase process). A practical specific capacity of 120 mAh/g has been demonstrated, but a loss of capacity of about 20% is sometimes observed on cycling. Since practical electrode configurations involve blends of $LiMn_2O_4$ powder with carbon black and a binder, this loss of capacity has been attributed to lack of interfacial contact and non-homogeneity in the electrode structure. Other interpretations include side reactions and/or dissolution in the electrolyte. Recently reported results suggest that improvement in capacity retention may be obtained by optimizing the morphology and blend ratios and by refining the synthesis procedures. The most common preparation procedure for $LiMn_2O_4$ is by reacting stoichiometric amounts of Li_2CO_3 and Mn_2O_3 in

Fig. 7.18 Schematic structure of $LiMn_2O_4$ spinel. (By permission of Dr A.R. Armstrong, University of St Andrews.)

air at 850°C. Optimization can be achieved either by controlling the cooling rate or by reducing the Mn^{3+} ion concentration. By developing new synthetic routes, including those based on sol-gel or solution methods, further improvement in cycle life has been achieved. Better cyclability has also been demonstrated following incorporation of small quantities of foreign atoms, such as Co, Ni or Cr in the $LiMn_2O_4$ structure. Recent reports suggest that high capacity, layered structure $LiMn_2O_4$ compounds can also be synthesized.

A novel $LiNiVO_4$ positive material, having a voltage of approximately 4.8 with respect to Li, has been characterized and proposed for high voltage lithium ion batteries. However, its practical use may be limited by the lack of electrolytes capable of withstanding its high oxidation potential. Other recent developments include NASICON-related framework structures, such as a $Fe_2(SO_4)_3$-based compound.

The final choice of the positive to be used in practical lithium ion batteries depends on the specific requirements of a particular developer. The characteristics of the three most common lithium metal oxide electrodes currently exploited in lithium ion technology are summarized in Table 7.3.

Although $LiCoO_2$ is still the most popular cathode in commercial lithium ion batteries, $LiNiO_2$ and, particularly, $LiMn_2O_4$ have advantages in energy density or cost. Issues of safety and rate limitation at low temperature may limit the use of $LiNiO_2$ in commercial batteries. Most of the new development projects are now considering the use of $LiMn_2O_4$ as the preferred positive material. Fig. 7.19 shows typical charge–discharge performance of a carbon/$LiMn_2O_4$ lithium ion cell; characteristics of commercial lithium ion cells using high voltage positives will be considered more fully in Section 7.5.

Table 7.3 Characteristics of 'high voltage' lithium metal oxide positives

Cathode	Capacity (Ah/g)	Average voltage (V vs Li)	Cost	Toxicity
$LiCoO_2$	0.180	3.9	Very high	High
$LiNiO_2$	0.220	3.8	Fair	Fair
$LiMn_2O_4$	0.120	4.0	Low	Low

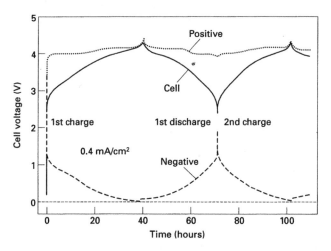

Fig. 7.19 Typical charge–discharge profile for a carbon–$LiMn_2O_4$ lithium ion cell cycled at 0.4 mA/cm². (By permission of *J. Electrochem. Soc.*: D. Guyomard and J.M. Tarascon, 1992, **139**, 937.)

7.4 Electrolytes

There are two groups of electrolytes currently in use or under development for ambient temperature, rechargeable lithium batteries. The first group are liquid solutions of lithium salts in aprotic organic solvents (usually solvent mixtures); the second group are known as polymer electrolytes and are formed by the dissolution of lithium salts in high molecular weight coordinating macromolecules such as poly(ethylene oxide) (PEO) or by incorporating them in a non-aqueous gel system.

The most commonly used electrolytes for lithium batteries are liquid solutions of lithium salts in aprotic organic solvents. As already discussed in Chapter 4, the main parameters which govern the choice of the electrolyte are:

- Solvent
  - stability towards both the positive and negative electrodes;

  – high solubility for lithium salts;
  – safety and non-toxicity.

• Lithium salt
  – high ionic mobility for the lithium ion;
  – thermal, electrochemical and chemical stability;
  – compatibility with the environment.

Solvents were initially selected primarily on the basis of the conductivity of their salt solutions, the classical example being propylene carbonate (PC). However, solutions based on PC on its own were soon found to cause poor cyclability of the lithium electrode, due to uncontrolled passivation phenomena. Solvent mixtures or blends were therefore developed and selection currently focuses on a combination of high dielectric solvents (e.g. ethylene carbonate, EC) with an alkyl carbonate (e.g. dimethylcarbonate, DMC), to stabilize the protective passivation film on the lithium electrode, and/or with a low viscosity solvent [e.g. 1,2-dimethoxyethane (DME) or methyl formate (MF)], to ensure adequate conductivity.

Other solvent combinations are found to optimize cell performance with carbon-based negatives. Considering the critical role of passivation phenomena, it is obvious that the response and cycling behaviour of carbon electrodes is greatly influenced by the nature of the passivation film, and consequently by the type and composition of the electrolyte solution selected. The effect on the capacity of a typical graphite host electrode of varying the solvent is shown in Fig. 7.20. The most popular choices for liquid electrolyte lithium ion batteries using graphite anodes are solutions of $LiPF_6$ or $LiN(CF_3SO_2)_2$ in EC–PC or EC–DMC solvent mixtures. In the

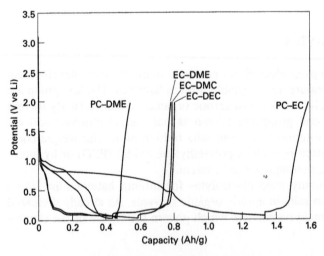

Fig. 7.20 Effect of the solvent on the capacity of C (graphite) negatives. (By permission of *J. Power Sources*: S. Megahed and B. Scrosati, 1994, **51**, 79.)

structurally disordered coke electrodes, the intercalation process does not promote formation of staged phases, and these electrodes are not as sensitive to the nature of the electrolyte as are the more crystalline graphite electrodes. For instance, PC-containing electrolytes, which readily degrade graphite electrodes, can be successfully used with coke electrodes.

The choice of salt is also important. The major considerations are related to thermal and electrochemical stability (which reflect on safety), toxicity and cost. Solutions using lithium perchlorate ($LiClO_4$), while moderately conductive, may be subject to violent decomposition, particularly in ether solutions. Lithium hexafluoroarsenate ($LiAsF_6$) suffers from thermal and electrochemical instability, which leads to formation of AsF_5 and insoluble LiF. Lithium hexafluorophosphate ($LiPF_6$) undergoes thermal decomposition at around 30°C in the solid state and 130°C in solution. Lithium trifluoromethanesulphonate ($LiSO_3CF_3$), while thermally and electrochemically stable, tends to be associated in solution, resulting in low conductivity. All of these problems have stimulated extensive research which has led to the introduction of new 'optimized' salts. Important examples are the bis(trifluoromethanesulphonyl) imide, $LiN(SO_2CF_3)_2$, and the tris(trifluoromethanesulphonyl) methide, $LiC(SO_2CF_3)_3$, which appear to have advantages over lithium salts of other anions, such as a good conductivity combined with a high stability. These favourable properties are related to the fact that both the imide and methide salts undergo little ion association, because of the significant charge delocalization. It is also important to note that the lithium transference number is generally higher in imide and methide solutions than in other salt solutions, due in part to the larger size and in part to the greater solvation of these anions. A high cation transference number leads to low concentration polarization, as noted in Chapter 2. One problem of the imide salt is its tendency to corrode aluminium, which is one of the most common positive current collectors in lithium batteries.

The lithium polymer battery (LPB), shown schematically in Fig. 7.21, is an all-solid-state system which in its most common form combines a lithium ion conducting polymer separator with two lithium-reversible electrodes. The key component of these LPBs is the polymer electrolyte and extensive work has been devoted to its development. A polymer electrolyte should have (1) a high ionic conductivity; (2) a lithium ion transport number approaching unity (to avoid concentration polarization); (3) negligible electronic conductivity; (4) high chemical and electrochemical stability with respect to the electrode materials; (5) good mechanical stability; (6) low cost and (7) a benign chemical composition.

The initial development of polymer electrolytes involved the formation of complexes between lithium salts such as $LiClO_4$, $LiCF_3SO_3$, $LiBF_4$ or $LiN(CF_3SO_2)_2$ and high molecular weight polymers containing Li^+-coordinating groups such as poly(ethylene oxide) (PEO). PEO–LiX electrolytes may be formed by a simple casting technique which involves the dissolution of PEO and the lithium salt in a suitable solvent (e.g. acetonitrile); the

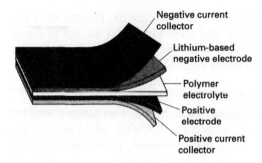

Fig. 7.21 Schematic illustration of the construction of a lithium polymer battery (LPB)

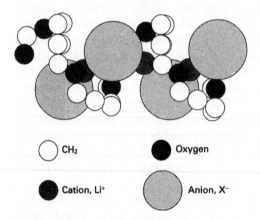

Fig. 7.22 Detail of typical crystalline polymer electrolyte phase showing PEO helix surrounding lithium ions

two solutions are then mixed and, after extended stirring, the solvent is slowly evaporated to form a film of thickness between 25 and 1000 μm. The formation of the electrolytes involves the coordination of the lithium ions by the oxygen atoms of the PEO polymer chains, with a basic structure involving the wrapping of the chains around the Li^+ cations in a helical structure. Crystalline phases may be obtained (e.g. Fig. 7.22) but these are not good ionic conductors. In all-amorphous materials above the glass transition temperature, local relaxation and segmental motion of the polymer host chains confer high conductivity on the electrolyte. Because of the tendency of the PEO to crystallize below 70°C, the conductivity of simple PEO–LiX electrolytes reaches practically useful values (i.e. of the order of 10^{-4} S/cm) only at temperatures above this (*see* Fig. 7.23).

To obtain improved ionic mobility, and thus high conductivity, alternative polymer structures have been developed, for example comb-branched block copolymers such as poly[bis(methoxy ethoxy ethoxide)], usually known as MEEP. Room temperature conductivities of MEEP-based polymer electrolytes of the order of 10^{-5} S/cm have been achieved, values

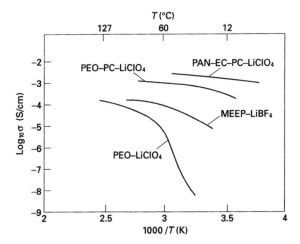

Fig. 7.23 Temperature variation of the conductivity of typical polymer electrolytes

which, however, are still two orders of magnitude lower than those of most liquid electrolytes. These types of polymer electrolyte have poor mechanical properties.

Further improvements in conductivity have been obtained by adding plasticizers [e.g. low molecular weight polyethyleneglycols (PEGs) or liquid aprotic solvents, such as propylene carbonate (PC) or ethylene carbonate (EC)] to the PEO–LiX electrolytes. The amount of plasticizer can be as high as 70%, but this results in limited mechanical properties and poor compatibility with the lithium metal negative. Another approach is to immobilize liquid solutions (e.g. solutions of lithium salt in a PC–EC organic solvent mixture) in a polymer matrix, such as poly(acrylonitrile), poly(methylmethacrylate) or poly(vinylidene fluoride). The immobilization procedure can be carried out by cross-linking the matrix using chemical methods, UV, electron-beam or gamma ray irradiation. 'Gel-type' polymer electrolytes of this sort have conductivities which approach those of liquids, namely 10^{-3} S/cm at room temperature.

The temperature dependence of the conductivity of the various classes of polymer electrolyte discussed above is summarized in the Arrhenius plots in Fig. 7.23. While a wide choice of materials is now available, it is important to note that improvements in conductivity are generally accompanied by losses in chemical stability and by increases in reactivity towards the lithium metal electrode. Successful development of rechargeable LPBs is therefore likely to be linked to the use of the so-called 'dry' polymer electrolytes, namely pure PEO–LiX systems. This necessarily confines the operation of LPBs to above ambient temperatures. This restriction does not apply to lithium ion cells.

7.5   Organic solvent-based practical cells

Cells with lithium metal negative plates

There have been a number of attempts to produce commercial lithium rechargeable batteries. The V_2O_5 positive is currently used by the Matsushita Battery Industrial Co in Japan for the production of small capacity, coin-type cells. Fig. 7.24 shows a cross-section of one prototype. For the construction of the battery, V_2O_5 and carbon black are mixed together with a binder, moulded and vacuum-dried to form the positive electrode pellet. A solution of $LiBF_4$ in a propylene carbonate-γ-butyrolactone-1,2-dimethoxyethane mixture absorbed in a polypropylene separator is used as the electrolyte.

Some details concerning this battery, which is designed for low drain loads, were listed in Table 7.2. The most important performance characteristics are the very low self-discharge (2% per year at 20°C) and long cycle life (over 1000 cycles at shallow depths of discharge). The energy density is, however, limited by the low capacity of the V_2O_5 polycrystalline cathodes. Current studies are aimed at the development of alternative V_2O_5 structures where a larger part of the intrinsic capacity of the oxide can be profitably exploited. Some promising results have been obtained with amorphous V_2O_5 analogues. Coin-type lithium batteries using V_2O_5–P_2O_5 vanadate glass positive electrodes have also been developed. These vitreous phases show high capacity (up to three lithium ions per mole at a 1.5 V cut-off), high energy density (900 Wh/kg) and long cycle life. More recently, coin cells using V_2O_5 xerogels have also been developed.

The earliest fully commercial rechargeable lithium battery was announced in the late 1980s by a Canadian company, Moli Energy Ltd.

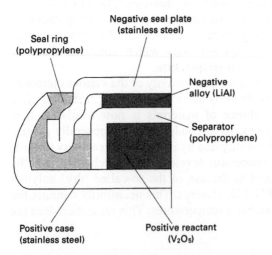

Fig. 7.24   Cross-section of a LiAl–V_2O_5 rechargeable coin cell. (By permission of Matsushita Battery Industrial Co.)

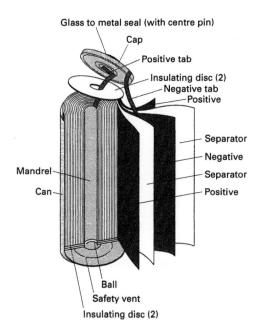

Glass to metal seal (with centre pin)

Cap

Positive tab

Insulating disc (2)
Negative tab
Positive

Separator

Negative

Separator

Positive

Mandrel

Can

Ball
Safety vent
Insulating disc (2)

Fig. 7.25 Construction of a Li–MoS$_2$ AA-sized rechargeable cell. (By permission of Moli Energy Ltd.)

This battery was based on a lithium metal anode and a molybdenum disulphide cathode:

$$Li(s)|LiAsF_6-PC(l)|MoS_2,C(s)$$

The electrochemical discharge–charge process of the battery is based on the intercalation/de-intercalation of lithium molybdenum disulphide:

$$x Li(s) + MoS_2(s) \underset{charge}{\overset{discharge}{\rightleftharpoons}} Li_x MoS_2(s) \qquad (7.12)$$

Cells were manufactured at AA size with a rated capacity of 0.6 Ah. Fig. 7.25 shows a cut-away drawing of a Molicel battery, and Fig. 7.26 shows a typical cycling performance. Despite a cycle life of over 400 and an energy density of 50 Wh/kg, safety hazards were identified which were associated with the lithium negative electrode, especially when the cell was abused.

Work has therefore been devoted by a number of developers to improving the cyclability of the lithium metal electrode. Since passivation of lithium is an unavoidable phenomenon, one approach has been directed to the promotion of uniform and smooth surface passivation layers, for example by selecting the most appropriate combination of solvents and electrolyte salts. An example is the inclusion of 2-methyltetrahydrofuran (2-Me-THF), since the presence of the methyl group slows down the reactivity towards the lithium metal. The selection of fluorine-based elec-

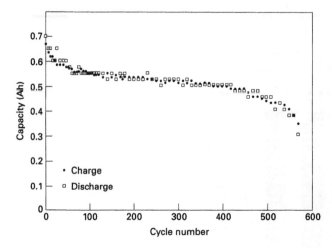

Fig. 7.26 Capacity of a Li–MoS$_2$ AA-sized rechargeable cell as a function of cycle number

trolyte salts, such as LiPF$_6$, has been shown to enhance the lithium cyclability, probably because of the formation of dense LiF-containing passivation layers. Substantial improvement in lithium cycling efficiency has also been obtained by using dialkylcarbonates, such as dimethylcarbonate (DMC) and diethylcarbonate (DEC). Further enhancement may be obtained by the addition of CO$_2$, whose beneficial effect is attributed to the formation of compact and high stable films of LiCO$_3$, HCO$_2$Li and ROCO$_2$Li. Another approach is to add small quantities of HF. Examination of the lithium surface after cycling the electrode both in standard solutions and in HF-added solutions shows that the presence of even small amounts of HF is very effective in promoting the passage from increasingly rough and irregular deposits to smooth and dendrite-free lithium deposits.

Finally, improvements in the reliability and the safety of lithium metal-based batteries have also been pursued by combining the use of optimized electrolyte solutions with external and internal safety mechanisms. The former are based on electronic circuits, outside the cell (generally in the upper part of its case), which prevent short-circuiting or polarity inversion. Internal safety measures may involve the use of 'shutdown' separators or of specially designed electrolytes, which spontaneously react to form products which rapidly deactivate the cell following local increases in temperature. These protect the cell from thermal runaway resulting from the reactivity of abnormally large lithium surfaces induced by dendrite growth following operational abuse.

This approach has been adopted by the Battery Division of Tadiran Ltd in Israel which is producing a rechargeable lithium battery under the trade name of 'Tadiran in-charge'. This battery, manufactured in a standard AA size, uses a lithium metal negative, a lithiated manganese oxide positive

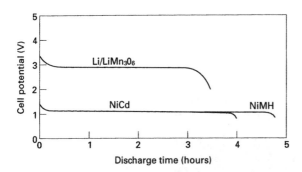

Fig. 7.27 Comparison of discharge behaviour of Li–LiMn₃O₆ AA-sized cells with nickel–cadmium and nickel–metal hydride cells. (By permission of Tadiran.)

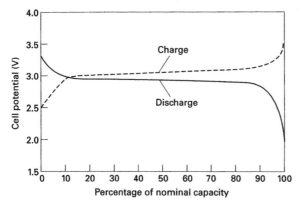

Fig. 7.28 Charge–discharge cycle behaviour of Li–LiMn₃O₆ AA-sized cells. (By permission of Tadiran.)

and a polypropylene porous separator which contains a solution of $LiAsF_6$ in 1,3-dioxolane (DN) stabilized with tributylamine (TBA):

$$Li(s)|LiAsF_6, DN-TBA(l)|LiMn_3O_6(s)$$

The main electrochemical reaction of this cell is

$$2Li(s) + LiMn_3O_6 \underset{charge}{\overset{discharge}{\rightleftharpoons}} 3LiMnO_2 \qquad (7.13)$$

The rated capacity is 0.75 Ah. Typical discharge curves and charge–discharge cycle behaviour of these 3 V batteries are shown in Figs 7.27 and 7.28, respectively. Energy density of 130 Wh/kg and 280 Wh/dm³ and 300 complete cycles at 100% depth of discharge have been reported. The electrolyte solution is stable at low to medium temperatures, but polymerizes when the temperature exceeds 125°C to form polydioxolane with a consequent rapid rise in the internal resistance of the cell. This provides an internal shut-off safety mechanism which, combined with a built-in vent,

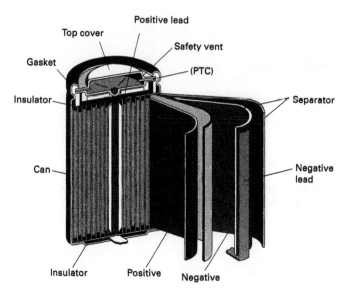

Fig. 7.29 Construction of a C(coke)–LiCoO$_2$ cylindrical rechargeable cell. (By permission of Sony Energytech.)

has allowed this cell to pass a stringent series of safety tests, including short-circuiting, overdischarge, crush and nail penetration.

Lithium ion cells

Japanese manufacturers have established a commanding lead in producing commercial lithium ion batteries. The first announcement was in February 1990, with the introduction to the consumer electronic market by the Sony Energytech Inc. of a camcorder lithium ion rechargeable battery. This battery was based on C(coke)/LiCoO$_2$ chemistry and had the cylindrical structure shown schematically in Fig. 7.29. The positive and negative composite electrode layers are applied uniformly to both sides of metal foil current collectors (typically copper for the negative and aluminium for the positive), and formed into a spiral. A microporous polymer separator contains the electrolyte solution. The cell package is housed in a cylindrical metallic case.

A number of safety devices are incorporated in this cell. The operation of the shut-off device is illustrated in Fig. 7.30; in the event of thermal runaway with rapid build-up of internal pressure, the (internal) positive electrode lead is disconnected, thus cutting off the current. Instructions on use and handling are included in the sale package of this and all commercial lithium ion batteries. In the case of the Sony product, these precautions include restrictions on operating temperature (typically not exceeding 45–50°C, especially during charge), discharge current (to be maintained below the 2 C rate), and charging and discharging voltage (4.20 and 2.5 V,

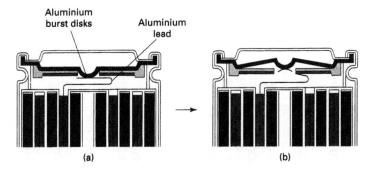

Fig. 7.30 Disconnection of the internal positive lead following a build-up of internal pressure in the Sony Energytech C(coke)–$LiCoO_2$ rechargeable cell, by permission. (a) Prior to operation. (b) After operation

respectively). Further, an external protection circuit is recommended during operation of either single cells or battery packs, designed to prevent overcharging and overdischarging, and excess current flow during cycling. Current production by Sony is approaching 10 million lithium ion cells per month by the middle of 1997. Both cylindrical and prismatic cells are manufactured. Fig. 7.31 shows charging characteristics at various current rates and Fig. 7.32 shows the capacity retention on cycling of prototype cylindrical cells (18 mm diameter × 650 mm height) designed to power video recorders. This cell has a working voltage of 3.6 V, an energy density of 125 Wh/kg and 300 Wh/dm^3, and a cycle life exceeding 500 cycles.

Table 7.4 shows a comparison of some lithium ion technologies currently under development. Rechargeable lithium ion batteries are now produced in configurations which range from small button cells to large cylindrical or prismatic sizes. The operating voltage, which is three times higher than that of nickel–cadmium cells, the relatively low self-discharge rate, the absence of the memory effect, the smooth slope of the discharge curve and the high energy density have led to increasing consumer interest and a consequent significant increase in manufacturing output. Projections indicate that by the beginning of the next century, the market share of secondary lithium ion batteries will reach at least 10% of the total worldwide sale of batteries for consumer electronics (Fig. 7.33).

Although the largest production of lithium ion batteries is directed towards the electronic consumer market, there are indications that scaling up to high capacity modules for electric vehicle use is also feasible. The LIBES Japanese project which is supporting the development of the lithium polymer battery (LPB), which is in the course of fabrication and test at Yuasa, is also sponsoring a large programme for EV lithium ion batteries involving a number of Japanese industrial companies. This focuses on two types of batteries, long-life and high energy density, which differ mainly in the target values of energy density and cycle life, viz. 120 Wh/kg,

Table 7.4 Comparison of lithium ion technologies being developed by various companies

Manufacturer	Anode material	Cathode material	Cell type	Cell size	Status
Sony, Japan	Coke, graphite	$LiCoO_2$	Cylindrical	AA, 14500 20500, 16630, 18650 14×34×48 13×33×48	Full production: 3.5 m/month (1996) 10 m/month (1997)
Panasonic Matsushita	Graphite	$LiCoO_2$	Cylindrical Prismatic	17500, 18650 8.3×40×48	Full production: 2 m/month Japan (1996) 4 m/month (1997)
Sanyo, Japan	Graphite	$LiCoO_2$	Cylindrical Prismatic	18650, 18500 8.1×22.5×48	Full production: 3 m/month (1996) 7 m/month (1997)
A&T Asahi Chemicals & Toshiba Corp Japan	Coke	$LiCoO_2$	Cylindrical Prismatic	18650 17670 8.6×34×48 6.3×30×48 14×34×48	Full production: 2.5 m/month (1996) 3.2 m/month (1997)
JS Japan Storage Japan			Prismatic	8.3×4×49 6.4×22×46	Production: 1.0 m/month (1996)
VARTA Germany	Graphite	$LiCoO_2$	Unknown	Unknown	Pilot line
SAFT France	Coke	$LiNiO_2$	Cylindrical	D	Production
Rayovac, USA	Coke	$LiNiO_2$	Coin Cylindrical	1225, 2335 AA, D	Pilot line Experimental

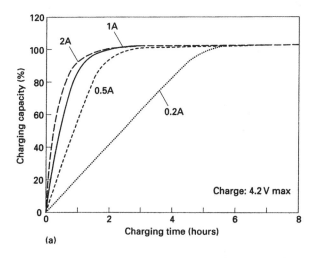

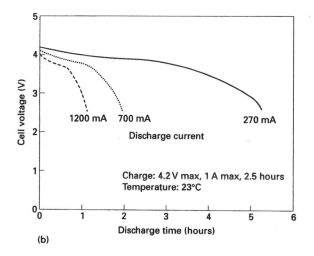

Fig. 7.31 Charging (a) and discharging (b) characteristics of the Sony Energytech C(coke)–LiCoO$_2$ rechargeable cell, by permission

240 Wh/dm^3 and 3500 cycles for the former, and 180 Wh/kg, 360 Wh/dm^3 and 500 cycles for the latter. The project has a three-phase programme: the first phase, already complete, involved the production of 10 Wh prototypes; the second, fixed for the year 1998, requires the scaling up to prototypes of 100 Wh; and the third, expected by the year 2001, is concerned with improvements in reliability of the 100 Wh modules. A list of the Japanese companies involved in the LIBES project and their chosen technologies is reported in Table 7.5.

The United States Advanced Battery Consortium (USABC) and the European Union are also presently supporting R&D projects for EV lithium

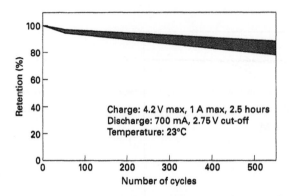

Fig. 7.32 Capacity retention of the Sony Energytech C(coke)–LiCoO$_2$ rechargeable cell, by permission

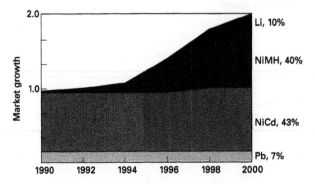

Fig. 7.33 Market projection for consumer electronics batteries

ion batteries. Further, a number of industrial laboratories are developing lithium ion EV prototypes independent of government sponsoring agencies. For example, the French company SAFT has announced the production of a complete EV battery, including thermal management and electronic control systems. Cells of 100 Ah capacity have been assembled and tested with 100 Wh/kg and 200 Wh/dm^3 performance. Mass production is expected from the year 2000. Even more advanced is the effort of the Japanese Sony Corporation where, as announced in 1995, it currently has under production and test a lithium ion EV battery having the performance and design specification summarized in Fig. 7.34.

7.6   Polymer-based practical cells

Lithium polymer batteries (LPB)

The LPB negative is commonly a lithium metal foil. The positive is based on a reversible intercalation compound, generally of the same type as those

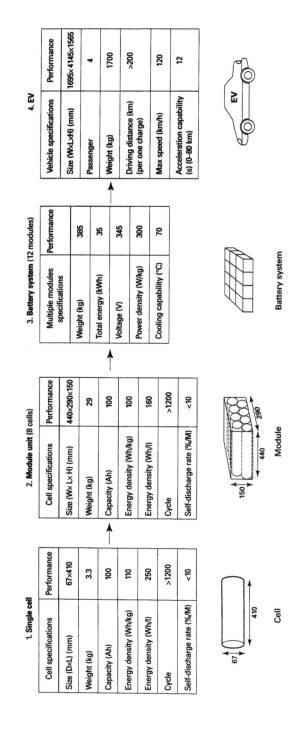

1. Single cell

Cell specifications	Performance
Size (D×L) (mm)	67×410
Weight (kg)	3.3
Capacity (Ah)	100
Energy density (Wh/kg)	110
Energy density (Wh/l)	250
Cycle	>1200
Self-discharge rate (%/M)	<10

Cell

2. Module unit (8 cells)

Cell specifications	Performance
Size (W×L×H) (mm)	440×290×150
Weight (kg)	29
Capacity (Ah)	100
Energy density (Wh/kg)	100
Energy density (Wh/l)	160
Cycle	>1200
Self-discharge rate (%/M)	<10

Module

3. Battery system (12 modules)

Multiple modules specifications	Performance
Weight (kg)	385
Total energy (kWh)	35
Voltage (V)	345
Power density (W/kg)	300
Cooling capability (°C)	70

Battery system

4. EV

Vehicle specifications	Performance
Size (W×L×H) (mm)	1695×4145×1565
Passenger	4
Weight (kg)	1700
Driving distance (km) (per one charge)	>200
Max speed (km/h)	120
Acceleration capability (s) (0–80 km)	12

EV

Fig. 7.34 Specification of the Sony Corporation lithium ion EV battery. (By permission of Sony Co.)

Table 7.5 R&D projects for EV lithium ion batteries in progress in Japan within the LIBES programme

Type	Company	Cell structure	Cathode	Anode	Electrolyte
Long-life	Mitsubishi Electric	Folding type	$LiCo_{0.3}Ni_{0.7}O_2$	Graphitized carbon	$LiClO_4$/EC+DME
	Hitachi	Prismatic	$LiNiO_2$	Ag dispersed graphite	$LiBF_4$/EC+DMC
	Sanyo	Cylindrical	$LiCo_{0.3}Ni_{0.7}O_2$	Natural graphite	$LiPF_4$/EC+DMC
	Yuasa	Prismatic	$LiCoO_2$	Graphite	$LiBF_4$/polymer
High-energy density	Japan Storage Battery	Prismatic	$LiCo_{0.95}B_{0.05}O_2$	Graphite	$LiPF_4$/ EC+DMC+DEC
	Matsushita	Cylindrical	$LiM_nCo_{0.2}O_4$	Ga-carbon	$LiPF_6$/EC+DMC
	Nippon Denso	Cylindrical	$LiMn_2O_4$	Ca dispersed Li metal	$LiPF_{67}$/EC+DME
	Toshiba	Prismatic	$LiNi_{0.97}B_{0.03}O_2$	Hard carbon	$LiPF_6$/EC+DMC

EC = ethylene carbonate; DME = dimethoxyethane; DMC = dimethylcarbonate; DEC = diethylcarbonate.
Derived from LIBES (Lithium Battery Energy Storage Technology Research Association) released information.

Fig. 7.35 The three component layers of an LPB (front to back): positive, polymer electrolyte and lithium foil

used for liquid electrolyte lithium battery systems (e.g. TiS_2, V_6O_{13}, LiV_3O_8 or $LiMn_2O_4$), as noted above. However, in the case of LPBs, the intercalation positive is blended with the PEO–LiX electrolyte and carbon to form a plastic composite which is backed by a metal foil current collector. The basic structure of an LPB can then be written as

$$Li(s)|PEO\text{–}LiX(s)|A_zB_y\text{–}PEO\ LiX,C(s)$$

Complete cells are formed by laminating the electrode and the electrolyte components together to obtain a flexible, compact structure with a total thickness of typically 200 μm. Fig. 7.35 shows the three films prior to lamination, while Fig. 7.36 is a microphotograph of a cross-section of the cell showing the three component layers. The electrolyte and the composite positive electrode can both be obtained using a range of coating techniques. For laboratory prototypes, 'doctor blade' solvent casting is common. However, the lamination process can be scaled up by using automatic winding machinery of a design somewhat similar to that routinely used in the capacitor industry.

A key advantage of the LPB concept is related to its all-solid-state construction. This facilitates the production of rugged (i.e. tolerant to shock, vibration and mechanical deformation), leak-proof, gassing-free and non-fixed geometry cells. Also, the absence of free liquids allows LPBs to be packaged in lightweight plastic containers, unlike conventional lithium batteries which require metallic casing. This simple but effective cell encapsulation is illustrated in Fig. 7.37 for a prototype cell.

Various geometries of LPBs, including a flat thin design for 'smart' credit cards, prismatic packaging for portable computers, C-size cylindrical cells for consumer electronic devices and high capacity cylindrical, spiral

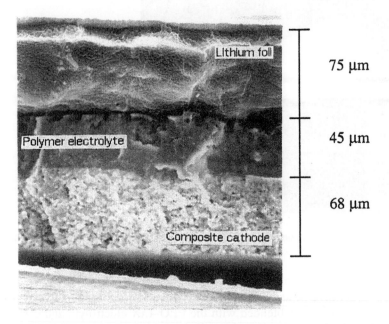

Lithium foil

75 μm

Polymer electrolyte

45 μm

68 μm

Composite cathode

Fig. 7.36 Cross-section of a typical LPB. Fractured edge of completed cell. Standard HT cell (40 cm^2) (magnification ×300) (By permission of AEA Technology.)

Fig. 7.37 Aluminium-coated plastic ('coffee bag') encapsulation of a laboratory prototype LPB

Fig. 7.38 Package designs for LPBs. (By permission of AEA Technology.)

assembly or prismatic stack models for the electric vehicle application, have been reported. Fig. 7.38 shows the various geometries of LPBs.

Most of the LPBs under development operate at around 100°C, i.e. at the temperature where the PEO–LiX electrolyte reaches useful conductivity. This is not a serious drawback if the battery is designed for the EV application where operational temperatures higher than ambient can be controlled by relatively simple thermal management techniques. However, high temperature operation excludes the use of poly(ethylene oxide)-based LPBs in the electronic consumer market. Research has therefore been devoted over recent years to the development of polymer electrolytes with enhanced transport properties at ambient and subambient temperatures. Highly conducting gel-type electrolytes have been characterized, but as discussed above in Section 7.3, these have poor compatibility with the lithium metal electrode which limits battery cyclability and may induce safety hazards. Therefore, in the large-scale development of LPBs, 'dry' PEO–LiX polymer electrolytes are still preferred.

Funding agencies, such as the European Union, the United States Advanced Battery Consortium (USABC), the Japanese Lithium Battery Energy Storage Technology Research Association (LIBES) and various government programmes, are presently heavily supporting LPB R&D projects for electric vehicle use. Demonstration prototypes have confirmed the competitive performance of LPB-based systems. Some examples are noted in Table 7.6.

Due to the competitive nature of the projects, it is difficult to report in any detail the progress achieved in the development of EV prototypes. For the majority of the cases, the preferred LPB positive electrode is a vana-

Table 7.6 Major R&D projects aimed towards the development of LPB prototypes for electric vehicle applications

Project	Sponsor	Electrolyte	Cathode	Status	Target
AEA Technology, UK	EU	PEO–LiClO$_4$	V$_6$O$_{13}$	10 Wh prototypes	Scaling-up
IREQ-3M-ANL, Canada-USA	USABC	M–PEO–LICLO$_4$ LiN(CF$_3$SO$_2$)$_2$	V$_6$O$_{13}$	10 Wh prototype	Scaling-up
Danionics, Denmark	EU	Gel-type	V$_6$O$_{13}$	Prototypes	Scaling-up
YUASA, Japan	LIBES	Polymer electrolyte	LiCoO$_2$	10 Wh prototypes	Scaling-up 20 kWh modules
CEREM France	Bolloré Techn EDF	PEO–LiX	VO$_x$	Prototypes	Scaling-up

EU = European Union; USABC = United States Battery Advanced Consortium; IREQ = Hydro-Quebec; ANL = Argonne National Laboratory; EDF = Electricité de France; LIBES = Lithium Battery Energy Storage Technology Research Association.

dium oxide, generally V_6O_{13}, although TiS_2 and $LiMn_2O_4$ have also been used. V_6O_{13} was the material selected in one of the earliest programmes, the 'Anglo-Danish project', which commenced in 1979 under EU sponsorship and continued at the Harwell Laboratory in the UK with the collaboration of Varta, Germany (until 1995), and the University of Rome, Italy. The cell developed was

$$Li(s)|PEO–LiX(s)|V_6O_{13}(composite)(s)$$

Test cells operating between 80 and 140°C having cathode capacity of 2.0–2.5 mAh/cm^2 range and initial energy density values approaching 200 Wh/kg were reported. Fig. 7.39 shows the cycle performance of one of the first versions of this cell. It is seen that 100% of theoretical capacity can be achieved in the first cycles; this falls, however, during extended cycling. This loss of capacity is typical of 'non-optimized' LPB cells and work is being undertaken at present to improve the cycling performance. Reports from Harwell suggest that no incidents have occurred when cells were overcharged and overdischarged, short-circuited, heated above 400°C and cut open. These experiments were run on test cells and the safety characteristics of EV units are still to be ascertained. Recent work has been directed towards scaling up the cell active area and the construction of 80 Ah units. The results obtained on these have highlighted the need for capacity balance.

Another project underway to develop large LPBs for electric traction which also dates back some years was started initially in Canada under IREQ-Hydro-Québec sponsorship and presently continues under a IREQ-3M-Argonne National Laboratory collaboration within the USABC framework. This project focuses on a technology using a lithium metal negative, a polymer electrolyte formed by a modified PEO copolymer containing $LiClO_4$ and $LiN(CF_3SO_2)_2$, together with a VO_x positive. Cells, having the structure

$$Li(s)|M-PEO–LiClO_4–LiN(CF_3SO_2)_2(s)|VO_x(composite)(s)$$

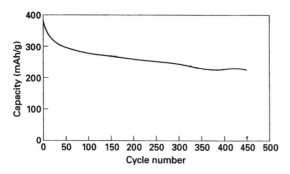

Fig. 7.39 Cycle performance of an early version of the 'Anglo-Danish' or Harwell/AEA Technology LPB, at 120°C and C/3. (By permission of AEA Technology.)

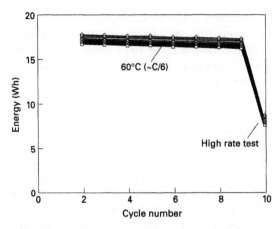

Fig. 7.40 Initial cycling performance of a group of LPB cells at C/6 from the IREQ-3M-ANL project. (By permission of Elsevier: M. Gauthier, *J. Power Sources*, 1995, **54**, 163.)

operate at a temperature of about 80°C. The use of a modified PEO (M-PEO) in combination with an optimized salt mixture, $LiClO_4$–$LiN(CF_3SO_2)_2$, permits the appreciable lower operations temperature of this LPB in comparison with that of cells using standard PEO electrolytes. A threefold excess of lithium compared to the stoichiometry of the positive electrode is used to compensate for losses during cycling.

A flat plate design involving double positive/central negative, bi-cell individual units has been selected for the construction of large capacity batteries. In this design the lithium foil is used as both active material and negative current collector; a thin aluminium foil is used as positive current collector. Winding the laminate into a spiral gives very compact, cylindrical structures. Demonstration 10 Wh prototypes have been fabricated and tested. The initial performance of these prototypes confirms 100% Ah efficiency and > 85% cycling energy efficiency by cycle 9 (Fig. 7.40). The next development will be the scale-up to EV modules, with a target of 200 Wh/kg specific energy.

The Danish company 'Danionics' has also developed LPB prototypes under EU contract using a lithium metal negative, a polymer gel electrolyte and a V_6O_{13} positive. The cells had the form

$$Li(s)|gel\text{-}electrolyte(s)|V_6O_{13}(composite)(s)$$

and operated in the temperature range of 0–60°C. Prototypes having capacities of 0.6 Ah and energy density values of 150 Wh/kg and 300 Wh/dm^3 were reported. The company are now concentrating on lithium ion-based systems which use polymer electrolytes (PLI).

Interest in LPBs is also fairly high in Japan where research is mainly focused on the development of new polymer structures having high conductivity and good mechanical strength. R&D projects for EV polymer batteries are currently in progress under the auspices of the LIBES programme. The Yuasa Corporation is involved in the production of LPB prototypes using a $LiCoO_2$ cathode. The LIBES final target, set for the year 2001, is the production of long-life batteries having an energy content of 120 Wh/kg and of 240 Wh/dm^3, and a life of 3500 cycles. Presently the Yuasa battery has met the phase 1 target, namely the construction and test of 10 Wh prototypes. Future phases will require module battery fabrication (by 1998) and reliability improvements.

Other LPB R&D projects have been launched in various countries throughout the world. In 1993, the Bolloré Technologies in collaboration with EDF and CEA/CEREM announced a programme focused again on a PEO–LiX polymer electrolyte and a vanadium oxide-based composite positive. The first objective was to demonstrate 40 Ah, 300 charge–discharge cycle prototypes and future plans will consider scaling up to EV modules with projected energy density of the order of 130 Wh/kg.

Work in progress at the Lawrence Berkeley Laboratory in California under the support of the US Department of Energy is focusing on LPBs using a polyorganodisulphide (PDS) positive

$$\text{Li(s)|PEO–LiX|PDS(composite)(s)}$$

The advantages of these organosulphur compounds, a typical example being poly-2-5-dimercapto-1,3,4-thiazole, are the long cyclability and the high energy density which they confer. Prototypes operating at 80–100°C have shown specific energies from 140 to 200 Wh/kg and from 170 to 240 Wh/dm^3 (depending on the discharge rate), and a life of several hundred cycles. There are still problems associated with the capacity fading on cycling.

A more recent LPB development in Israel involves the fabrication of bipolar cells using a composite polymer electrolyte (CPE), and a pyrite-based positive

$$\text{Li(s)|CPE|FeS}_2\text{–PEO(s)}$$

The composite electrolyte is formed by a mixture of LiI, PEO and Al_2O_3. The addition of alumina improves the stability of the lithium electrode. Results reported by the University of Tel Aviv suggest that prototype Li/FeS_2 cells operating at 130°C have an impressive projected energy density, viz. 250 Wh/kg. This high value, together with the low cost of the positive material, makes the system of considerable interest. Problems associated with this technology are related to cyclability and capacity retention.

Scale-up from laboratory test cells to EV module is the next challenge for the LPB technology. There are three general areas which need to be addressed when considering scale-up, namely: (1) raw materials, (2) component fabrication, and (3) cell and battery construction. In general, the raw materials employed in the various forms of lithium polymer batteries can easily be obtained in large quantities. The key areas are the lithium metal foil and the active positive material. Lithium metal foils are commercially available in a range of thicknesses down to 50 μm. However, thinner

(< 30 μm) and wider (> 10 cm) foils will not be easy to produce in reasonable quantities and at low cost. Scale-up of the various active positive materials employed in LPBs is a matter of concern, since, particularly with vanadium-based positives, the cell performance appears critically dependent on the reproducibility of the production processes employed to fabricate the active material.

The scale-up of cell components is a matter of adapting well-known, large-scale industrial processes to suit the special needs of batteries. Specifically, it is important that the components are fabricated in a dry atmosphere and that the addition of chemicals, to improve the coating quality, is carefully controlled. Several groups are known to be developing the required level of machinery to scale up cell components. In Canada, IREQ has produced components at a level required for a significant number of 10 Wh cells. This technology is now under expansion and improvement within the framework of an agreement with 3M and the Argonne National Laboratory. In the UK, the Harwell Laboratory has production machinery capable of delivering several kilometres of polymer electrolyte and composite positive. In the United States, Valence Technology have equipment capable of automatically producing lithium polymer battery components and cell laminate. The Danish company Danionics continues to expand its operation under EU directives. Scaling up lithium polymer batteries should be a simple matter of using larger areas of laminate. Producing the cell laminate from individual components is again a task in which existing industrial techniques may be adapted. Therefore, one would assume that performance obtained with test cells can be reproduced with larger size cells. However, data proving that the polymer batteries can maintain performance when scaled up to factors of 1000 (or even 100) are not yet available and it is in the demonstration of this possibility that most of the laboratories are presently involved.

7.7   Plastic lithium ion (PLI) batteries

An important development in lithium ion battery technology involves the replacement of the liquid electrolyte with a polymer electrolyte to produce what has been termed the plastic lithium ion battery (PLI). This is an interesting concept, since it provides the prospect of a favourable combination of the high energy and the long cycle life typical of the lithium ion cell, with the reliability and processability typical of a polymer-based all-solid-state configuration. The practical exploitation of PLIs requires the availability of polymer electrolytes having a conductivity approaching that of the liquid solutions, as is the case for LPBs described above. Gel-type polymer electrolytes formed by trapping liquid solutions in a polymer network described in Section 7.3 are being exploited. Other requirements for the successful development of practical PLIs are: the fabrication of electrode films which can combine flexibility and high loading capacity with good adherence to the current collector and low electronic resistance; and compatibility between the electrode and the electrolyte film components. The former is essential to ensure high rate and high energy capabilities, and the latter to ensure long cyclability and high capacity delivery. The second requirement appears to be the most challenging. While the fabrication of highly loaded electrode films with good mechanical properties is now seen as straightforward in various industrial laboratories, a satisfactory inter-layer contact between these electrodes and the electrolyte in a practical solid-state cell configuration is not easily achieved, especially at the negative, carbon electrode side. The major problem is to ensure a full wettability throughout the entire electrode mass, a process which is naturally achieved when the electrode is immersed into a liquid solution but which must be 'activated' or contrived in a different way when the same electrolyte is trapped in a polymer matrix.

A successful approach has been reported by the Bellcore Laboratory in the United States with the use of a membrane separator formed as a copolymer of vinylidene fluoride and hexafluoropropylene (PVDF–HFP), which is capable of absorbing large quantities of liquid electrolyte. The role of the HFP component is to decrease the crystallinity of the PVDF copolymer and thus to enhance its ability to absorb liquid. By optimizing its composition, the PVDF–HFP membrane can absorb a liquid electrolyte to approximately 200% of its solid volume while still retaining its good mechanical properties. In a further modification, the polymer membrane is processed by adding highly dispersed fillers and is formed using a plasticizer which is subsequently extracted and replaced by the electrolyte. This modification produces PVDF–HFP copolymers consisting of an optimized amorphous and crystalline phase distribution; while the amorphous regions hold large amounts of liquid electrolyte, thus ensuring high ionic conductivity, the crystalline domains provide the mechanical strength and processability.

The basis of the Bellcore PLI battery technology involves a continuous production process in which a stack consisting of two electrode layers (a carbon-based negative and a $LiMn_2O_4$ positive) separated by the plasticized electrolyte membrane is fused by lamination to form a unified tri-layer strip. Next, the plasticizer is removed by extraction with an organic solvent. The cell laminate is then housed in an aluminium-coated bag or other container, freed from solvent by reduced pressure and/or elevated temperature, and activated by injection of the desired amount of the selected electrolyte solution, generally a solution of $LiPF_6$ in EC: DMC at a ratio of 2:1. Thermal sealing of the package concludes the fabrication process.

This technique ensures good interfacial contact between the layers which in turn leads to low internal resistance (and thus to high rate capability) and to long cycle life. A cyclability of up to 1500 cycles to a cut-off of 80% of initial capacity at the 1 C rate at room temperature, energy densities of the

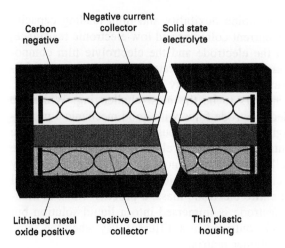

Fig. 7.41 Schematic cross-section of a PLI rechargeable cell. (By permission of Ultralife Batteries Inc.)

order of 130 Wh/kg and 300 Wh/dm^3, and costs comparable if not less than the liquid electrolyte counterparts have been reported.

Ultralife Batteries Inc. has recently announced the development of their Rechargeable Solid State System™ based on this technology. A schematic drawing of the cell is shown in Fig. 7.41. Single cells and battery packs are now in production. Energy densities in the ranges 100–125 Wh/kg and 200–300 Wh/dm^3, as well as cyclability up to 1200 deep cycles, have been reported. Excellent characteristics are claimed. Cellular telephones, portable computers and camcorders appear to be the most suitable outlets for these PLI batteries.

8 High temperature cells

J L Sudworth

8.1 Introduction

The primary requirements for large-scale energy storage batteries, designed for applications such as vehicle traction, load levelling, etc., are high power and energy densities, coupled with low cost. For optimum performance and commercial success, electrode materials must therefore be selected which are very reactive and also abundant in nature. This in turn leads to the conclusion that lithium and sodium, being among the lightest, most electropositive and abundant of metals, should be ideal anode materials for such advanced battery systems. Because of their reactivity, these alkali metals cannot be used in conjunction with aqueous electrolytes. Two further classes of electrolyte have therefore been exploited, namely fused salts and solid ionic conductors. Systems using fused salt electrolytes must be maintained at temperatures high enough to keep the electrolyte in a molten condition; batteries with solid electrolytes must be operated at temperatures which assure a sufficiently high conductance of the electrolyte, and that in general the electrode phases are in a liquid condition to minimize electrode/electrolyte contact problems. One can therefore define a new group of power sources, which, requiring operating temperatures above ambient, are classified as high temperature cells.

Many problems concerning reliability, cycle life, safety and others connected with the advanced technology of these systems have been solved. The remaining challenges are to establish the process required for mass production. Production has reached a level of hundreds of batteries per year for some types of high temperature batteries, i.e. tens of thousands of cells. Additionally, very considerable research and development programmes are being devoted to these batteries and it is considered likely that one or more systems will make a significant commercial impact in the near future. Before describing the construction and performance of individual systems, the basic properties of lithium and sodium electrodes and electrolytes commonly used in high temperature batteries will be briefly reviewed.

8.2   Negative electrodes

Lithium-based electrodes

The physical properties of lithium metal were given in Table 4.4. Despite its obvious attractions as an electrode material, there are severe practical problems associated with its use in liquid form at high temperatures. These are mainly related to the corrosion of supporting materials and containers, pressure build-up and the consequent safety implications. Such difficulties were experienced in the early development of lithium high temperature cells and led to the replacement of pure lithium by lithium alloys, which despite their lower thermodynamic potential remained solid at the temperature of operation and were thus much easier to use.

Of the various solid intermetallic lithium compounds which might be used in high temperature cells, the Li–Al system has been most studied. The Li–Al phase diagram is shown in Fig. 8.1. An α phase which consists

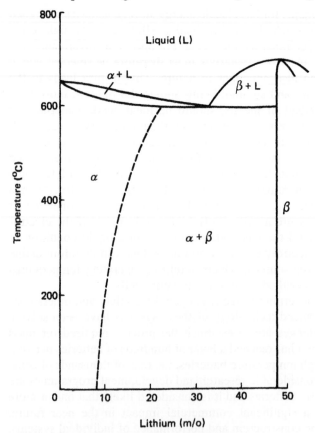

Fig. 8.1 The lithium–aluminium phase diagram

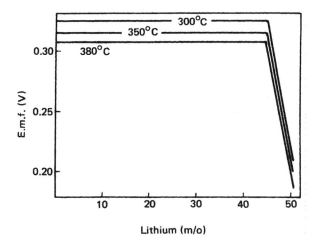

Fig. 8.2 Emf of lithium–aluminium alloy as a function of temperature and composition. (By permission of *J. Electrochem. Soc.*: N.P. Yao *et al.*, 1971, **118**, 1039.)

of a solid solution of lithium in aluminium is stable up to about 7–9 atoms per cent of lithium depending on the temperature. With increasing lithium content, a β-phase or Li–Al compound is formed. The α–β two phase region extends to about 46 atoms per cent of lithium.

The emf of the lithium–aluminium system versus pure lithium in a LiI–KI–LiCl molten eutectic is shown in Fig. 8.2 as a function of temperature and composition. It can be seen that the emf remains constant (at about 300 mV more negative than pure lithium) in the range of stability of the β-phase (≈7–47 atoms per cent of lithium), thus implying a constant lithium activity in the alloy surface. At concentrations greater than 47 atoms per cent, the lithium activity becomes strongly composition-dependent.

Li–Al alloys may be prepared electrochemically, generally by coulometric deposition of lithium from a molten salt bath, or pyrometallurgically by heating lithium and aluminium at a temperature above the alloy melting point of 720°C. Practical electrode configurations are constructed by:

- electrochemical deposition on a substrate of compressed aluminium fibres;

- hot-pressing powdered mixtures of pyrometallurgical Li–Al and electrolyte; or

- loading pyrometallurgical Li–Al powder into a porous nickel mass using vibratory techniques.

All these operations, and more generally the assembly of the complete cell, must be performed in a controlled, water-free atmosphere.

Li–Al electrodes behave well in molten salt electrolytes such as the LiCl–KCl eutectic, showing low polarization and good reversibility, with

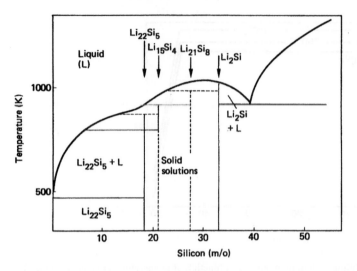

Fig. 8.3 The lithium–silicon phase diagram. (By permission of the *Journal of the Electrochemical Society*: R.A. Sharma and R.N. Seefurth, 1976, **123**, 1763.)

flat charge/discharge characteristics, even under high current densities. Cycling behaviour may, however, be affected by progressive capacity losses. This problem may be partially solved by further modification of the alloy by the addition of a ternary component. Inclusion of approximately 4% by weight of indium significantly increases capacity retention on cycling.

The Li–Si system has also been investigated as a possible high temperature solid lithium electrode. The lithium–silicon phase diagram, illustrated in Fig. 8.3 reveals five different compounds in the lithium-rich region. The existence of these compounds may be demonstrated electrochemically by the coulometric titration of a silicon electrode in a LiCl–KCl eutectic molten electrolyte, as shown in Fig. 8.4. Typical charge/discharge cycles of Li–Si in the same electrolyte at 407°C (Fig. 8.5) show reversibility and good current efficiency, but also high polarization during charge. This latter effect, possibly due to restricted lithium permeation within the short composition ranges of some of the phases, has so far restricted the use of Li-Si alloys in practical high temperature rechargeable batteries.

Sodium electrodes

Sodium is also a very reactive metal, and with a melting point even lower than that of lithium, presents in principle problems similar to those of lithium. However, the fortunate discovery of ceramic materials which show high stability to molten sodium together with good sodium ionic conductivity at high temperature has permitted the reliable fabrication of sodium-based cells. In some sodium high temperature cells, the liquid metal is housed in closed, shaped ceramic containers. In the others, the

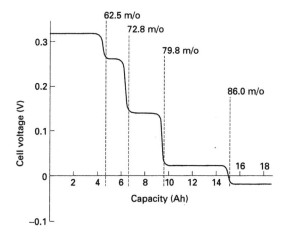

Fig. 8.4 Coulometric titration of a silicon electrode with lithium using a lithium counter electrode and a LiCl–KCl eutectic at 680 K. Current density = 1 mA/cm². (By permission of the *Journal of the Electrochemical Society*: as Fig. 8.3.)

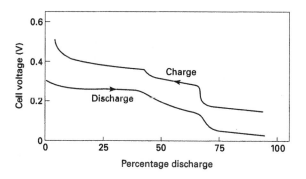

Fig. 8.5 Typical charge–discharge characteristics of a lithium–silicon electrode in a LiCl–KCl eutectic at 680 K. Current density = 40 mA/cm². (By permission of the *Journal of the Electrochemical Society*: as Fig 8.3.)

sodium is held in steel cases or chambers. Special assemblies are constructed for vacuum-filling cells with molten sodium to prevent contact with oxygen or moisture. Alternatively, cells can be assembled in the discharged state and the sodium metal generated on the first charge.

8.3 Electrolytes

Lithium molten electrolytes

Molten single salts or molten eutectic mixtures are generally characterized by very high ionic conductivities (> 100 S/m). This property makes them desirable media as battery electrolytes since it permits operation at high

Table 8.1 Physical properties of some lithium and sodium molten electrolytes*

Salt(s)	Composition	M.P. (°C)	Density (g/cm^3)	Viscosity (cP†)	Conductance (S/m)
LiCl		610	1.490 (637°C)	1.377 (637°C)	585.4 (637°C)
LiCl–KCl	58.5 m/o LiCl 41.5 m/o KCl	355	1.646 (447°C)	1.46 (617°C)	161.5 (457°C)
LiF–LiCl–LiBr	22 m/o LiF 21 m/o LiCl 47 m/o LiBr	445	2.190 (500°C)	—	—
NaCl–KCl	50 m/o NaCl 50 m/o KCl	685	1.571 (717°C)	1.58 (727°C)	239.6 (717°C)
NaCl–AlCl$_3$	50 m/o NaCl 50 m/o AlCl$_3$	155	1.691 (177°C)	2.645 (187°C)	46.2 (187°C)

* From G. Mamantov in 'Materials for Advanced Batteries', D.W. Murphy, J. Broadhead and B.C.H. Steele, Editors, Plenum Press, N.Y., London, 1980, p. 113.
† 1 cP = 0.001 kg m^{-1}s^{-1}

current densities with low iR drop, and hence without undue heat dissipation problems. On the other hand, molten salt electrolytes require the maintenance of high temperatures. In addition to the necessity of heating the cell (at least initially) this requirement leads to a series of severe technical problems concerned with corrosion of cell housings, separators, ceramics for feedthrough insulators, current collectors and seals. The most common electrolyte used in lithium high temperature batteries is the LiCl–KCl eutectic which has a lower melting point than single lithium halides, but has the requisite high ionic conductance. However, pure lithium dissolves in this electrolyte to an extent of about 0.13 m/o at 400°C, which may lead to some self-discharge of cells. In addition, metallic lithium reacts with KCl to form potassium vapour with an equilibrium partial pressure of 0.05 kPa at 425°C. The properties of some lithium molten electrolytes are given in Table 8.1.

Molten chloroaluminates

Chloroaluminates, which are $AlCl_3$–MCl mixtures (where M is an alkali metal), offer a number of advantages as molten electrolytes, such as low melting point, high conductance and excellent stability (when contamination with oxygen and moisture is avoided). This melt can be considered as an acid–base system with the relative acidity or basicity being related to the chloride ion concentration. This in turn is governed by various equilibria; the predominant process near to the 1:1 $AlCl_3$/MCl ratio is

$$2AlCl_4^- \rightleftharpoons Al_2Cl_7^- + Cl^- \tag{8.1}$$

The $AlCl_3$ – NaCl mixtures have been used in various high temperature batteries. The physical properties of the 1:1 system are summarized in Table 8.1.

Sodium β-aluminas

Sodium β-aluminas (often known as simply 'β-aluminas') are examples of solid electrolytes, i.e. compounds which permit fast ionic motion (here of sodium ions) within a solid lattice. While β-aluminas conduct reasonably well at room temperature (≈ 1 S/m for polycrystalline material), they are generally used at temperatures over 300°C where their conductance is greater than 10 S/m. ('Ambient' solid electrolytes and batteries based on these will be considered in the next chapter.)

The most common phase is β-alumina itself, which has a formal composition of $Na_2O.11Al_2O_3$ but which in practice always contains an excess of Na_2O. Some of the physical properties of this material are given in Table 8.2. The high sodium ion mobility can be understood by examining the hexagonal layered β-alumina structure, shown schematically in Fig. 8.6. The structure is seen to contain planes having loose-packed sodium and oxygen ions. Sodium planes are held 1.13 nm apart by four close-packed

Table 8.2 Physical properties of β- and β''-aluminas

	β-alumina	β''-alumina
Melting point	2000°C	—
Density (at 25°C)	3.25 g/cm^3	—
a lattice constant	0.559 nm	0.559 nm
c lattice constant	2.253 nm	3.340 nm
Ionic conductance		
Single crystal (at 25°C)	3.5 S/m	—
Single crystal (at 300°C)	21.5 S/m	100 S/m
Polycrystalline (at 25°C)	1.0 S/m	—
Polycrystalline (at 300°C)	6.5 S/m	20 S/m
Electronic conductance (at 300°C)	~10^{-9} S/m	~10^{-9} S/m

spinel blocks (oxygen layers with aluminium ions in octahedral and tetrahedral positions) which extend normal to the c-axis. The spinel blocks above and below the sodium planes are mirror images of each other and are separated by oxygen bridges of 0.48 nm. The sodium ions can move easily within the loosely packed planes, but not through the closely packed spinel blocks. The structure therefore allows a two-dimensional diffusion of the sodium ions in directions perpendicular to the c-axis.

Another important phase in the $Na_2O–Al_2O_3$ system with a similar layered structure is the β'' form, of formal composition, $Na_2O(5.33Al_2O_3)$, whose basic properties are also summarized in Table 8.2. This phase is stabilized by small additions of MgO or Li_2O. The level of stabilizer addition affects the properties of the material, and the optimum composition in terms of conductance and mechanical strength appears to be around 0.8% Li_2O (with Na_2O at 9%). β''-alumina has a higher conductance than β-alumina and is therefore preferable as a battery electrolyte. However, it is more sensitive to moisture and is thus more difficult to handle than β-alumina.

β-aluminas are always used as sintered polycrystalline masses because of the need to fabricate the electrolyte in particular shapes for battery application. The conductance of polycrystalline samples is lower than that of correctly oriented single crystals since the migration of sodium ions is affected by the tortuosity of the two-dimensional pathways and by grain boundary impedances. Indeed, at low temperature the sodium ion mobility is completely dominated by grain boundary effects, as is shown in Fig. 8.7 where the general trends in the conductivity of polycrystalline and single crystal β-aluminas as functions of temperature are compared.

Just as the electrical properties are clearly related to the phase composition and microstructure of the electrolyte, so are its key mechanical proper-

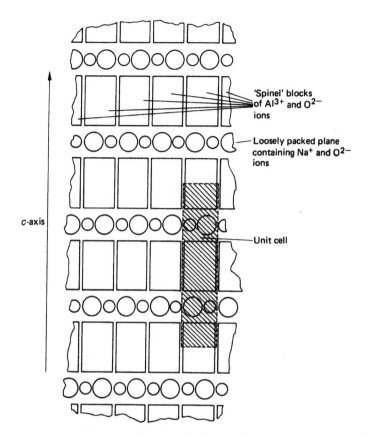

Fig. 8.6 Schematic diagram of β-alumina showing loosely packed planes containing mobile sodium ions situated between spinel blocks of Al^{3+} and O^{2-} ions

ties such as durability, resistance to thermal and mechanical shock, etc. Major research and development programmes are under way in Japan and the UK, to optimize electrical and mechanical properties and to fabricate the electrolyte in a suitable form. The most common configuration is of sintered, polycrystalline material in the form of a closed tube generally of diameter 2–3 cm, length 20–70 cm and wall thickness 1–2 mm. Some commercial battery electrolyte tubes are shown in Fig. 8.8.

Other high temperature solid electrolytes

Other crystalline sodium ion conductors, the so-called NASICON compounds of general formula $Na_{1+x}Si_xZr_2P_{3-x}O_{12}$ for $1.8 < x < 2.4$, have been investigated as possible alternatives to the β-aluminas. The advantage of this group of materials is their comparative ease of preparation.

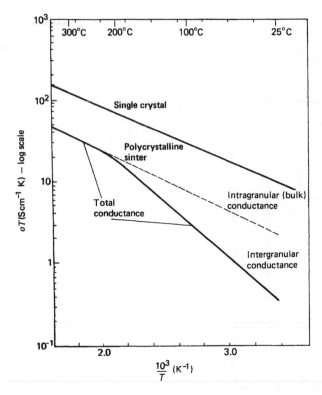

Fig. 8.7 Arrhenius plots for single crystal and polycrystalline β-alumina. The overall conductance of the latter is dominated by the intergranular impedance at low temperatures. (By permission of IOP: A. Hooper, *J. Physics*, 1977, **10**, 1493.)

Glass electrolytes

Sodium borate glasses which show some sodium ion conductivity ($\approx 4 \times 10^{-3}$ S/m at 300°C) have been known for many years. A sodium glass electrolyte using bundles of thousands of hollow glass fibres was used as an electrolyte in a sodium sulphur cell, but difficulties with sealing, corrosion of the glass and safety issues prevented further development. This configuration combines a very high superficial area with a wall thickness of some 10 μm, so that the total conductance of the electrolyte is high, despite the relatively high specific resistance.

From a technological point of view, there are many advantages in developing highly conducting vitreous materials for battery electrolytes. First, the effect of grain boundaries can often be eliminated and, in addition, it is often easier and cheaper to fabricate glass into the form required for the final battery system. A wide range of vitreous lithium conductors has been studied, in particular lithium-rich phosphate, borate and aluminate systems. Some of these are stable in contact with lithium, but no commercial cell has yet been reported using such an electrolyte. A range of

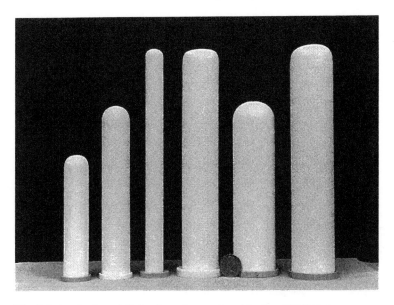

Fig. 8.8 A selection of β-alumina tubes produced by Beta R&D Ltd

sodium-conducting glasses is also known, with some Na_2S–SiS_2 glasses having as high an ionic conductance as the β-aluminas. Again, use of these glasses in commercial sodium-based cells has yet to be reported.

8.4　High temperature lithium batteries

Lithium–halogen and lithium–chalcogen batteries have a certain historical importance in the field of high temperature batteries, since they were among the first systems to reach a practical stage of development. However, severe problems were encountered, associated with the corrosion of the cell housing, seals and ancilliary components, and thus with the stability and safety of the cells, so that further development of these systems has now practically ceased. Nevertheless, the results obtained drew attention to the energetic possibilities of lithium high temperature cells in general, and thus stimulated further research on alternative stable lithium systems. It therefore seems useful to outline briefly the basic features of representative lithium–halogen and lithium–chalcogen batteries.

Lithium–chlorine batteries

The cell, shown schematically in Fig. 8.9, has the form

$$Li(l)|LiCl(l)|Cl_2(g),C(s)$$

The liquid lithium anode is held in a stainless steel case or nickel fibre pleated wick, and chlorine gas is fed under pressure from external storage

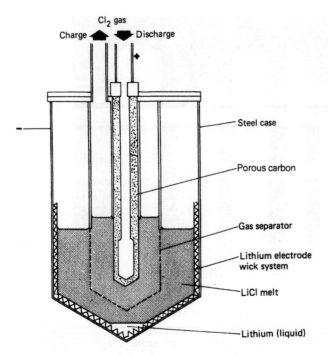

Fig. 8.9 Schematic diagram of a vertical experimental lithium–chlorine cell

to a porous carbon current collector. The two electrodes are separated by a molten lithium chloride electrolyte at 650°C. The products formed during recharge (i.e. liquid lithium and gaseous chlorine) must be liberated in such a way as to be easily separated and stored, in order to ensure proper cycling behaviour. This is achieved by using separator screens or special electrode structures which have porous, electronically insulating layers on the sides facing the electrolyte. When wetted by the electrolyte these layers form a seal which prevents or at least limits the escape of the products. The overall cell reaction is

$$\text{Li(l)} + \tfrac{1}{2}\text{Cl}_2(\text{g}) \rightarrow \text{LiCl(l)} \tag{8.2}$$

with which is associated an emf of 3.46 V and a theoretical energy density of 2.18 kWh/kg at the temperature of operation. However the practical realization of this very high energy is limited by a number of factors. Despite initial power densities of up to 40 W/cm^3, the available energy is reduced by the necessity of maintaining the cell at 650°C, and pumping the chlorine gas under pressure into the porous positive electrode. In addition, the 'valve' layers on the electrodes, described above, have a significant resistance which causes increased iR drop and thus depresses cell performance.

The most serious problems with this system are, however, concerned with corrosion of cell components and the development of satisfactory seals. Other approaches, such as the use of a closed system (based on the absorption of chlorine *in situ* on high surface carbon cathodes) and the reduction of operating temperature (by using the lower melting LiCl–KCl eutectic), have not proved successful, and there is little further development in progress at present.

Lithium–sulphur batteries

A typical example of this system might use the ternary eutectic LiF–LiCl–LiI, which melts at 341°C, and a cell operating temperature in the range 350–400°C. The cell would then have the form

$$Li(l)|LiF–LiCl–LiI(l)|S(l)$$

and the discharge process might be represented as

$$xLi(l) + S(l) \rightarrow Li_xS(diss.) \tag{8.3}$$

The average emf associated with this process is approximately 2.3 V and the theoretical energy density is 2.6 kWh/kg at 350°C.

⸱ In the version developed at the Argonne National Laboratories in the USA, the liquid lithium anode was contained in a stainless steel cup and was separated from the liquid sulphur cathode by a sheet of zirconia cloth. However the all-liquid system presented difficult practical problems, and a new type of cell using an immobilized electrolyte was proposed, where a more effective separation between the electrodes could be achieved. The electrolyte consisted of a rigid 'paste' of $LiAlO_2$ filler and LiF–LiCl–LiI eutectic in a 1:1 ratio. The first example of this type of cell had a tellurium cathode and was able to provide an emf of 1.7 V at 480°C and a practical energy density of 150 Wh/kg. However, once again problems of corrosion and thermal stress in cell components reduced the cycle life to a level below that acceptable for a practical storage battery.

Lithium alloy–metal sulphide high temperature batteries

The substitution of a solid alloy for liquid lithium and a transition metal sulphide for liquid sulphur, while lowering the energy density of the system, leads to a significant reduction in many of the operating problems associated with the early lithium–chalcogen cells and to easier fabrication of practical batteries. In particular, problems connected with the high vapour pressure of sulphur at the cell operating temperature (77 kPa at 425°C) and sulphur-displacement processes are eliminated, and the characteristic low polarization and good reversibility of the lithium alloy electrodes can be exploited. The theoretical energy density is reduced from 2600 Wh/kg to 400 Wh/kg, depending on the exact chemical composition of the cell. A wide variety of sulphides have been considered for the positive electrode, including those of Co, Cr, Cu, Mn, Ni, Ti and V, but FeS and FeS_2 have been selected for cell development on the basis of performance, cost and availability. Commercial batteries are being developed for EV applications mainly at the Argonne National Laboratory and SAFT America. 20 kWh batteries have been tested but these Li–Al/FeS systems were not suitable for high power/energy ratio electric vehicles such as passenger cars.

The cell being developed for low power/energy ratio applications is denoted as

$$LiAl(s)|LiCl–KCl(l)|FeS(s)$$

The cell reaction involves a two-electron process and may be written formally as

$$2LiAl(s) + FeS(s) \underset{charge}{\overset{discharge}{\rightleftharpoons}} Li_2S(s) + 2Al(s) + Fe(s) \qquad (8.4)$$

The emf of this reaction is 1.33 V and the corresponding theoretical energy density is 460 Wh/kg. In reality the actual mechanism of the cell reaction is very complex and not yet fully understood: it may involve as many as six electrochemical steps and four chemical reactions, and has a different sequence depending on whether the surface (i.e. the area in contact with the electrolyte) or the interior of the FeS particles is involved. The discharge reactions are thought to occur in the following sequence:

- Surface:

$$6LiAl + 26FeS + 6KCl \rightarrow LiK_6Fe_{24}S_{26}Cl + 2Fe + 5LiCl + 6Al \,(8.5)$$

- Interior:

$$2LiAl + 2FeS \rightarrow Li_2FeS_2 + Fe + 2Al \qquad (8.6)$$

$$2LiAl + Li_2FeS_2 \rightarrow 2Li_2S + Fe + 2Al \qquad (8.7)$$

- Surface:

$$46LiAl + LiK_6Fe_{24}S_{26}Cl + 5LiCl \rightarrow 26Li_2S + 24Fe + 6KCl + 46Al \quad (8.8)$$

The charging process is simpler and involves mainly the reactions

$$Fe + 2Al + 2Li_2S \rightarrow Li_2FeS_2 + 2LiAl \qquad (8.9)$$

and

$$Fe + 2Al + Li_2FeS_2 \rightarrow 2FeS + 2LiAl \qquad (8.10)$$

The formation of the compound $LiK_6Fe_{24}S_{26}Cl$ (which is also known as a mineral occurring in meteorites) appears to hinder the kinetics of the FeS electrode. The extent of the formation of this phase is dependent upon the quantity of potassium ions in the electrolyte, and a change from eutectic to lithium-rich electrolyte compositions, for either the LiCl–KCl or the LiCl–LiBr–Kbr electrolyte, eliminates or significantly suppresses the formation

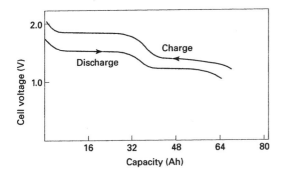

Fig. 8.10 Charge–discharge characteristics for a Li–FeS cell at 450°C. Battery assembled in uncharged state. Current = 5A. (By permission of Academic Press: W.J. Walsh and H. Shimotake, *Power Sources* 6, ed. D.H. Collins, 1977, London.)

of this phase. Typical charge/discharge curves are shown for this system in Fig. 8.10.

The cell

$$LiAl(s)|LiCl–KCl(l)|FeS_2(s)$$

has a higher emf (1.76 V) and theoretical energy density (650 Wh/kg) and is being developed for EV applications. The overall cell reaction is a four-electron process which may be written as:

$$4LiAl(s) + FeS_2(s) \underset{charge}{\overset{discharge}{\rightleftharpoons}} 2Li_2S(s) + 4Al(s) + Fe(s) \quad (8.11)$$

Again the actual cell reactions are very complex and the compositions of some of the ternary intermediates are only known approximately. The discharge mechanism is considered to proceed in three main steps:

$$3LiAl + 2FeS_2 \rightarrow Li_3Fe_2S_4 + 3Al \quad (8.12)$$

$$LiAl + Li_3Fe_2S_4 \rightarrow \text{various phases} \rightarrow 2Li_2FeS_2 + 2Al \quad (8.13)$$

$$2LiAl + Li_2FeS_2 \rightarrow 2Li_2S + Fe + 2Al \quad (8.14)$$

Traces of potassium compounds are also found due to reactions with the molten electrolyte. The sequence of phase changes during the charging process are even more complex. Typical charge/discharge curves are shown in Fig. 8.11.

Cell materials

As with all high temperature batteries, the materials problem in the lithium alloy–metal sulphide system is one of trying to develop low cost components which are able to withstand the hostile environment of the cell. The choice of insulators and separators is greatly restricted by the high stability of Li_2O which excludes use of the common ceramics such as Al_2O_3 and

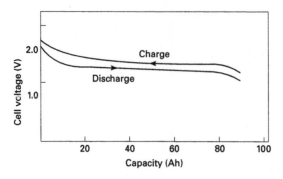

Fig. 8.11 Charge–discharge characteristics for a Li–FeS$_2$ cell at 450°C. Battery assembled in uncharged state. Current = 5A. (By permission of Academic Press: as Fig. 8.10.)

SiO$_2$ for thermodynamic reasons. Other materials have been found unsuitable because of impurity content, formation of conductive surface layers, etc. The only effective stable fabrics so far reported have been made of BN (boron nitride) or Y$_2$O$_3$ (yttria). The former, although very expensive, is some six times cheaper than the latter. BN fabrics have been used as separators for most of the Argonne National Laboratory development programme. A less expensive material is magnesium oxide powder which is mixed with the powdered electrolyte and pressed to form MgO/electrolyte plaques. These contain 70–80% electrolyte. Cell housings and current collectors are constructed of low-carbon steel where possible, i.e. in the FeS cells; expensive current collectors based on molybdenum alloys have proved necessary in FeS$_2$ cells. The electrodes consist of a porous bed of the solid active material located in an expanded metal or 'honeycomb' current collecting core, and flooded with molten salt. They are surrounded by a fine porous particle retainer or screen which confines the active material within the electrode structure. Electrodes may be formed in either the charged or discharged state.

Cell design and performance

The earliest prototype cells were cylindrical in form. A central metal sulphide electrode was embedded in a molybdenum mesh current collector and this electrode was surrounded first by a layer of ZrO$_2$ felt, acting as particle retainer, and then by a BN fabric separator. Upper and lower LiAl electrodes used the cell housing as current collector. Early cells of this design produced by the Argonne National Laboratory, called 'bicells' because of the double negative electrode, had typical practical energy densities in the range 100–140 Wh/kg but poor cycle life and capacity retention. However, the main thrust of lithium–iron sulphide cell design has been directed towards the production of vertically oriented cells with rectangular cross-section (Fig. 8.12). Again the central positive electrode is sandwiched between two negative electrodes which are in contact with the

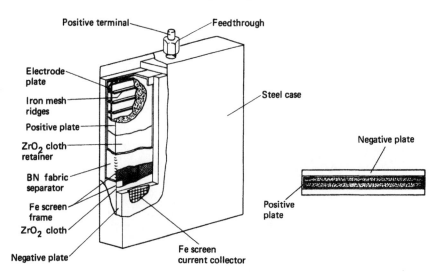

Fig. 8.12 Rectangular (prismatic) lithium–iron sulphide battery. (By permission of Academic Press: as Fig. 8.10.)

walls of the cell casing. The positive lead passes through an insulating compression seal in the top of the cell. Fine metal screens and/or zirconia or yttria felt are used as particle retainers and the positive mixture is wrapped in boron nitride fabric, which acts as separator. Typical Argonne National Laboratory cells are 13 cm × 13 cm or 13 cm × 18 cm and 1–3 cm thick, with theoretical capacities of 120–150 Ah. FeS-based cells have been shown to operate successfully to over 100 deep cycles, but with poor specific power; FeS_2-based cells, on the other hand, show peak specific power of more than 170 W/kg and practical specific energies of over 100 Wh/kg. The Eagle–Picher FeS cell is 4 cm thick, but has a multiplate design with three positive and four negative electrodes (in parallel) in the one housing. This unit is able to develop a specific power of 95 W/kg. A 40 kWh prototype EV battery has been constructed, consisting of two 20 kWh modules (Fig. 8.13). Each module contains 60 multiplate cells housed in a thermally insulated casing equipped with a vacuum insulation annulus to minimize heat losses.

The better performance of the FeS_2 electrode cells vis à vis the FeS cells led to these becoming the focus of the lithium–aluminium/iron sulphide battery development programme at Argonne. In 1986 a life of 1000 cycles was achieved on prismatic cells with flooded electrolyte and dense electrodes operating on the upper voltage plateau.

A problem which had made the operation of batteries difficult was the tendency for cells to get out of balance during cycling, presumably due to small differences in the rate of capacity loss. In 1988 this problem was overcome by the addition of Al_5Fe_2 to the negative electrode. This has the effect of increasing the lithium metal activity at the top of charge, thus providing a self discharge mechanism, i.e. migration of lithium dissolved

Fig. 8.13 20 kWh lithium–aluminium/iron sulphide battery. (By courtesy of Argonne National Laboratory.)

in the electrolyte to the positive electrode where it reacts chemically. Overcharge can then occur at a rate equal to or lower than the self-discharge rate (2–3 mA/cm^2). A side-effect of this is that boron nitride separators cannot be used. Instead, magnesium oxide compacted powder separators are used. The use of a starved electrolyte cell, necessitated by the magnesium oxide separator, required the development of a lithium-rich LiCl–LiBr–KBr electrolyte with a higher conductivity than the eutectic electrolyte. A change in the cell design from prismatic cell to bipolar cells was made possible by the development in 1990 of a new class of ceramic materials which can be engineered to be electrical insulators or conductors. These materials can be bonded to ceramics, metals and graphite. Typical performance data are given in Table 8.3.

Table 8.3 Specific energy and power of 13 cm diameter bipolar cells

Cell technology	Specific energy (Wh/kg at W/kg)	Specific power at 80% DOD* (W/kg)
Li–Al/FeS	130 at 25	240
Li–Al/FeS$_2$	180 at 30	400

*DOD = depth of discharge.

8.5   High temperature sodium batteries

Sodium melts at 98°C and therefore many of the materials problems experienced with the handling of liquid lithium might be expected to occur in the development of high temperature sodium batteries. However, the

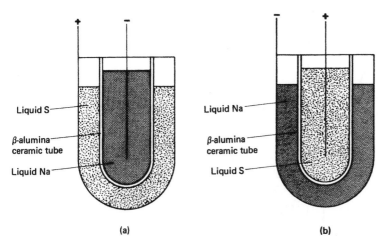

Fig. 8.14 Schematic diagram of the two basic cell configurations used in beta batteries

discovery of solid sodium ion conductors, especially β-alumina, revolutionized the design and fabrication of high temperature cells and has led to a massive research programme into a new type of battery technology over the past 30 years. The main interest up to 1979 was in sodium–sulphur batteries. Many groups in Europe, USA and Japan contributed to the development of this battery, but since 1995 only Yuasa Batteries, NGK and Hitachi in Japan are still working actively. In 1979 work began in South Africa on sodium/metal chloride batteries with alumina electrolyte and this battery is now under development in the UK and Germany as well as South Africa.

Sodium–sulphur batteries with β-alumina electrolyte ('beta batteries')

As shown in Fig. 8.14 the cell is formed in principle by two liquid electrodes, the sodium negative and the sulphur positive, separated by a tube of sintered polycrystalline β-alumina. Since sulphur is an insulator, the compartment containing the sulphur electrode is fitted with a carbon felt current collector. The cell, which may be written as

$$Na(l)|\beta\text{-alumina}(s)|S(l),C(s)$$

is operated in the temperature range 300–400°C where the reactants and discharge products are in the liquid state and the ionic conductivity of the β-alumina tube is high. As the cell is discharged, sodium atoms ionize at the surface of the β-alumina tube and migrate through the tube wall to form sodium polysulphides. The initial discharge reaction is

$$2Na(l) + 5S(l) \rightarrow Na_2S_5(l) \qquad (8.15)$$

which has an emf of 2.08 V at 350°C and a theoretical energy density of

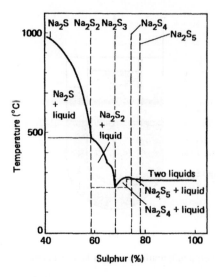

Fig. 8.15 The sodium sulphide–sulphur phase diagram. (By permission of N. Webber and J.T. Kummer, *Proceedings of the 21st Annual Power Sources Conference*, 1967.)

790 Wh/kg. As indicated in the sodium–sulphur phase diagram given in Fig. 8.15, sodium pentasulphide and sulphur are not mutually soluble at the temperature of cell operation, so that two liquid phases are present in the cathode compartment and the cell voltage is invariant. As the discharge progresses and the available elemental sulphur is consumed, a series of reactions commences as the sodium pentasulphide is converted to lower polysulphides, all of which are mutually soluble:

$$2x\text{Na(l)} + (5 - x)\text{Na}_2\text{S}_5\text{(l)} \rightarrow 5\text{Na}_2\text{S}_{5-x}\text{(l)} \quad \text{for } x \leq 2 \quad (8.16)$$

The cell voltage falls progressively from 2.08 to 1.78 V as the composition of the catholyte changes from Na_2S_5 to Na_2S_3 (Fig. 8.15). The normal working range does not extend beyond Na_2S_3 since if Na_2S_2 and Na_2S were formed, they could crystallize out.

As with all high temperature systems, sodium–sulphur cells must be heated up before use. However, once they have reached working temperature, iR losses (on discharge or charge) together with $T\Delta S$ contributions during discharge are sufficient to maintain them at the required temperature. In fact cooling is required under some operating conditions. In EV applications, thermal insulation together with only a very small self-discharge through an internal heater can keep an open circuit sodium–sulphur battery within the operational temperature range for several days.

Cell materials

Almost all practical sodium–sulphur cells are based on electrolytes formed as closed tubes. These are usually manufactured by isostatic pressing, or electrophoretic deposition of powdered β-aluminas (or their precursors) followed by batch sintering or continuous zone-sintering at 1600–1700°C. General Electric developed a successful electrophoretic forming technique to produce a very uniform green tube and this has been adopted by RWE Silent Power. The reliability of the electrolyte tube is the key to the success of the sodium–sulphur battery and a great deal of effort is being made to develop dependable units. As can be understood, many of the details of the electrolyte formation process are industrial secrets. One of the most difficult technical problems in the history of this system has been the development of seals to isolate the cell compartments from each other and from the external environment. Generally an insulating β-alumina header is first attached to the β-alumina tube by a glass seal. Aluminoborate glass has been shown to be superior for this purpose to the more commonly used silicate glasses. The β-alumina then requires to be bonded to the metal casings; a number of techniques have been employed but thermocompression bonding using an aluminium interlayer is almost universally used. Another materials science problem is concerned with the corrosion of the case or the central current collector by molten sodium polysulphides; various coating procedures and materials have been studied. If an aluminium cell case is used, the inside can may be coated with nichrome by flame or plasma spraying. If a mild steel cell case is used this can be chromized.

Cell design and performance

In all cells being developed commercially, the sodium is placed inside the electrolyte tube. This permits the use of smaller diameter tubes and a higher energy density for the cell.

As the discharge proceeds, sodium is removed from the anode chamber, but to ensure a low resistance at the sodium/electrolyte interface it is necessary to keep the entire surface wetted with sodium throughout the discharge.

This can be achieved in two ways: by capillary action using a wick or by a pressurized cartridge inserted in the β-alumina tube. If a wick is used it is normally a metal foil which can withstand heating to around 1000°C, as it is inserted before the glass sealing operation. If a pressurized cartridge is used the pressure can be provided by decomposition of sodium azide (at 300°C) which produces sodium metal and nitrogen gas. In both designs most of the sodium is contained within a metal container which has a small hole at the bottom. This is a safety feature which limits the amount of sodium immediately available for reaction with the sulphur electrode and restricts the rate at which the remaining sodium can react. The design of practical sodium–sulphur cells is illustrated in Figs 8.16 and 8.17. Typical charge/discharge curves for Na/S cells with a central sodium electrode after 30 cycles are shown in Fig. 8.18.

For EV applications (e.g. urban delivery vans) the target figures for a commercially viable sodium–sulphur battery would be for a cycle life of

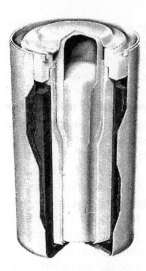

Fig. 8.16 Schematic diagram of a sodium–sulphur cell for electric vehicle applications

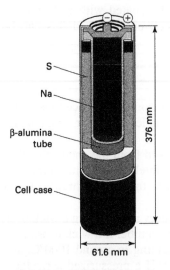

Fig. 8.17 Schematic diagram of a sodium–sulphur cell for load levelling applications

1000 cycles (i.e. 5 years continuous use) and an overall battery practical energy density of 100 Wh/kg. In addition, the design must incorporate adequate safety measures to minimize the effects of a collision or other accident. For load levelling use, requirements are not so strict, but cycle energy efficiency must be high.

Pilot plant manufacture is now in operation so that large-scale testing of batteries can be undertaken and a great deal of experience has been gained with laboratory sized cells, full engineering scale cells and commercial prototypes in the 50–200 Ah capacity range. Cell cycle lives of up to 1500

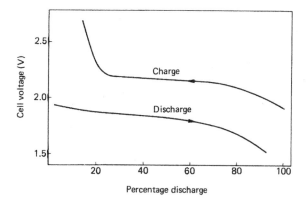

Fig. 8.18 Charge–discharge characteristics for a sodium–sulphur cell. (By permission of Elsevier Sequoia: F.G.R. Zobel, *J. Power Sources*, 1978, **3**, 29.)

Fig. 8.19 Bedford van powered by a sodium–sulphur battery

cycles have been reported and cell energy densities of 140 Wh/kg have been achieved. A number of traction batteries have already been built and tested, with energies of up to 60 kWh by RWE Silent Power and ASEA Brown Boveri. The 7.5 tonne van used for road testing sodium–sulphur batteries by Chloride Silent Power is shown in Fig. 8.19. The battery configuration for this vehicle consists of 30 parallel arrays of 100 cells in series with interconnections every four cells. Road tests have indicated a range of up to 160 km between charges.

A fleet of Ford 'Ecostar' vans is now operating in the US and Europe, powered by 34 kWh sodium–sulphur batteries. There have been a number of incidents in which sodium–sulphur batteries have been destroyed by the exothermic reaction which occurs when sodium and sulphur react chemically. Despite this, field trials are continuing indicating that these problems may not be insurmountable.

The sodium–antimony trichloride cell

The problems with the volatility and insulating nature of sulphur and the high melting points of the sodium polysulphides persuaded Werth to investigate a cell using β-alumina as the solid electrolyte and sodium as the negative electrode, but with antimony trichloride dissolved in sodium tetrachloroaluminate ($NaAlCl_4$) as the positive electrode. The operating temperature of the cell was 210°C and the overall cell reaction was

$$3Na(l) + SbCl_3(l) \rightleftharpoons 3NaCl(diss) + Sb(s) \qquad (8.17)$$

The theoretical specific energy of this reaction is 752 Wh/kg and the open circuit voltage should be 2.90 V. In practice the cell voltage is slightly higher in a fully charged cell, possibly due to the formation of $SbCl_5$.

Cells are assembled in the fully charged state and the positive electrode mix contains antimony trichloride (29%), sodium chloride (15%), aluminium chloride (48%), carbon black (8%) and small amounts of sulphur, which are added to give a higher average discharge voltage.

Although the boiling point of $SbCl_3$ is 219°C, the vapour pressure of the melt corresponding to the above composition is only about 600 Pa, indicating complex formation.

Because of the low operating temperature, PTFE can be used for the cell seals which simplifies cell assembly. The high operating voltage, however, leads to problems of corrosion and the only current collector material which has been found to show any corrosion resistance is tungsten. If sulphur is not added to the melt, a nickel current collector can be used, but omission of sulphur adversely affects the operation of the positive electrode. The problems encountered with the positive electrode were so severe that the development of this cell was discontinued.

The Zebra battery

One of the problems encountered with the Werth cell was an increase in resistance with cycling. This may have been caused in part by the β-alumina reacting with the acidic sodium chloroaluminate melt. Coetzer had the idea of using transition metal chlorides as a positive electrode and chose a basic sodium chloroaluminate melt as the liquid electrolyte. This is compatible with β-alumina, and a new class of secondary cells based upon the reaction between sodium metal and transition metal chloride has resulted from this work. Collectively, the term Zebra battery is used to describe this new class of cell.

The Zebra battery is presently at the pilot line production stage. About 170 batteries are on test, mostly in vehicles but some on bench tests. The batteries are installed in passenger cars and in buses, and in the German electric vehicle fleet test on the island of Rugen, 52% of the 60 or so vehicles are equipped with Zebra batteries. A lifetime of 4 years, 1260

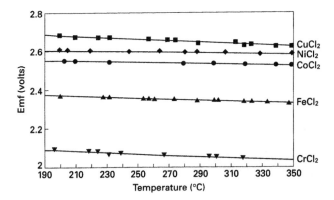

Fig. 8.20 Variation of emf with temperature for Na/β-alumina/NaAlCl₄/metal chloride cells

cycles and 110 000 km has been achieved in vehicle tests without the replacement of any cells.

Operating principles

Several of the first row transition metal chlorides offer attractive voltages versus sodium in the cell arrangement:

$$Na(l)|\beta\text{-alumina}(s)|NaAlCl_4,MCl(l)$$

and Fig. 8.20 shows the experimentally determined variations of open circuit voltage versus temperature for a number of these. Both $FeCl_2$ and $NiCl_2$ have been extensively developed as electrode materials. $FeCl_2$, which gives an OCV of 2.35 V, has the advantage of low cost but the cell chemistry is rather more complex. $NiCl_2$ has the advantage of a higher open circuit voltage (2.58 V), which for a given internal resistance leads to higher maximum power and for this reason has been favoured for EV application with its high power demand. The theoretical specific energy is 750 Wh/kg.

The cell reactions are shown in Fig. 8.21, from which it can be seen that overdischarge and overcharge reactions are possible. Use is made of the overdischarge reaction to allow the assembly of the cell without the addition of elemental sodium. To the products of the discharge reaction – nickel and sodium chloride – is added a small quantity of aluminium which on the first charge generates enough excess sodium to prime the sodium electrode:

$$Al + 4NaCl \rightleftharpoons NaAlCl_4 + 3Na \qquad E^\ominus_{cell} = 1.58 \text{ V} \qquad (8.18)$$

The main cell reaction involves chlorination of high surface area nickel powder with sodium chloride to form the nickel chloride positive electrode and sodium:

$$Ni + 2NaCl \rightleftharpoons NiCl_2 + 2Na \qquad E^\ominus_{cell} = 2.58 \text{ V} \qquad (8.19)$$

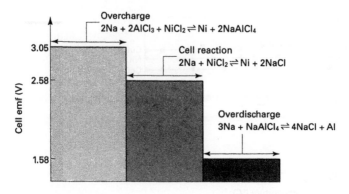

Fig. 8.21 Cell reactions

Sufficient sodium chloride is added to chlorinate over 30% of the nickel, after which the overcharge reaction occurs in which sodium chloride is abstracted from $NaAlCl_4$ to give an $AlCl_3$-rich melt.

Cell design and battery configuration

The cell design is shown schematically in Fig. 8.22. Two nickel components are sealed to a metallized β-alumina collar by thermocompression bonding; the β-alumina tube is then attached to the alumina collar by a glass seal. This sub-assembly can then be welded to the prismatic metal can. A metal wick inserted in the steel can, before the sub-assembly is welded in place, acts as the cathode for sodium deposition when the cell is charged. The positive electrode is added as a mixture of nickel powder, salt and aluminium powder, and after addition of the molten sodium chloroaluminate the cell is sealed by welding on the positive terminal

Every cell is tested by charging and discharging before assembly into a battery. This is done coulometrically so that all cells have the same capacity to within $\pm$ 0.5%. The criteria for acceptance are: top of charge voltage, discharge resistance and bottom of discharge voltage. The reproducibility of the cells is shown in Fig. 8.23 where the distribution of top of charge voltage is plotted. The accurate control of capacity combined with the shape of the discharge curve (Fig. 8.24) gives protection against cells being driven negative in batteries. This can be understood by considering the effects of one cell failure in a series/parallel connected battery. The lower voltage plateaus in Fig. 8.24, which probably correspond to reactions involving sulphur (added in small quantities to the positive electrode to stabilize the capacity), cover several ampere hours of capacity. Thus, in a series connected chain which continues to discharge, owing to circulating currents arising from imbalance, all the cells will discharge onto the lower voltage plateaus and the current will attenuate to zero thus preventing the cells being driven negative.

Batteries can operate with many failed cells because such cells have a

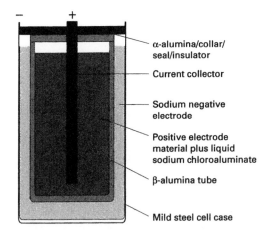

Fig. 8.22 Schematic design of a Zebra battery

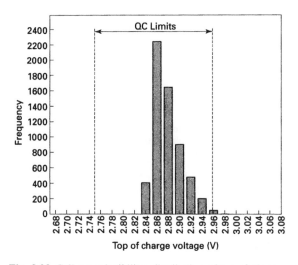

Fig. 8.23 Cell reproducibility: distribution of top of charge voltage

resistance comparable to operating cells. For this reason long series chains can be used in battery construction. Typical connection schemes are shown in Fig. 8.25, both of which have been used in batteries supplied to customers. The Z5 battery is one of a pair of batteries installed in each of the Mercedes Benz 190 cars which are operating on the island of Rugen.

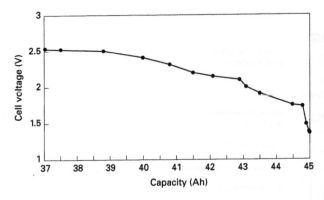

Fig. 8.24 Shape of end of discharge curve

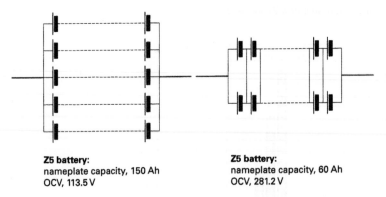

Z5 battery:
nameplate capacity, 150 Ah
OCV, 113.5 V

Z5 battery:
nameplate capacity, 60 Ah
OCV, 281.2 V

Fig. 8.25 Typical cell connections schemes: (a) Z5 battery – nameplate capacity, 150 Ah; OCV, 113.5 V; (b) Z5 battery – nameplate capacity, 60 Ah; OCV, 281.2 V

Battery design and characteristics

In common with other high temperature batteries, it is necessary to thermally insulate the battery to minimize heat loss. This is achieved by the use of a double-skinned stainless steel box in which 20–30 mm of insulation board is sandwiched between the two skins. Evacuation of this insulation ensures a low heat loss, and experience has shown that this vacuum is maintained over several years.

Because the positive electrode is contained inside the β-alumina tube, it is possible to use a cell case of square cross-section. This gives maximum packing efficiency and battery sizes can be changed in increments of 37 mm (cell width plus insulation) to meet customers' requirements. The battery is heated to the operating temperature by means of resistance heaters and these are also used to maintain the battery at the minimum operating temperature. A cooling system is provided for high rate discharge

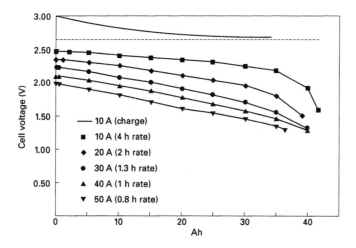

Fig. 8.26 Discharge performance at various rates

and this can use either air or a liquid which is stable to temperatures up to 400°C.

Heating and cooling are automatically controlled by a microprocessor-based battery controller which has the following functions: control of battery temperature; charge control and integration for state of charge display; electrical insulation resistance monitoring; voltage control; and fault detection and recording.

The specific energy of a Zebra battery suitable for powering electric cars approaches 100 Wh/kg and 160 Wh/dm^3, a value which is high enough to meet the required daily range of a city car. The specific power is typically around 150 W/kg and 250 W/dm^3 for a fully charged battery. Due to the increase in internal resistance during discharge, this power decreases. The capacity of the battery is virtually independent of discharge rate, as shown in Fig. 8.26 for a 10-cell module.

A life of at least 5 years and 1000 cycles is required for electric vehicle applications, and the Zebra battery has the potential to meet this. To date, no wear-out mechanisms have been identified: significant corrosion does not occur and after 1000 cycles the battery capacity is still greater than 80% of the nameplate capacity. One of the longest lived batteries is a 450 V, 13.5 KWh battery which was on test for 3 years. The capacity and resistance are shown as a function of cycles in Fig. 8.27. The battery was fully discharged each cycle and charged at 2.5 kW (equivalent to the power of an on-board charger). No cell failures occurred during this test. Other batteries have been tested on bench and in electric cars, and a summary of the test data is given in Table 8.4. The performance of these converted Mercedes Benz 190 cars is shown in Fig. 8.28.

Zebra batteries have been subjected to a series of tests to demonstrate their ruggedness and safety. These include overcharge, short circuit, overheating and vibration and shock. Drop testing to simulate the effect of a

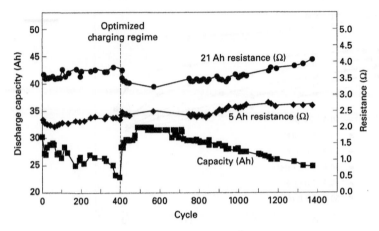

Fig. 8.27 Capacity and resistance vs. cycles for a 450 V, 13.5 kW battery

Table 8.4 Zebra batteries

Test	Battery type	Performance	Time on test (years)
Bench ECE 15	Z4	1700 cycles	2.3
Bench ECE 15	Z5	831 cycles	2.0
Bench DST	Z6	700 cycles	1.5
Bench 2h rate: d/c, normal charge	Z1	1640 cycles	3.3
Car DB	Z1	112 016 km	4.0
Car DB (Rügen)	Z5	> 100 000 km	3.5

severe road accident is an important feature of the battery qualification procedure and two types of test are used. The first test is to drop batteries hot and fully charged on to a 1 m length of 'Armco' crash barrier standing upright (Fig. 8.29). This penetrates the battery to a depth of two-thirds its width. Some white smoke is emitted from the battery and the external temperature is increased locally. In the second test a battery is dropped onto a semi-cylindrical steel pole, resulting in considerable distortion of the battery box. In both tests the terminal velocity is 50 km/h. These tests showed that there would be no danger to the driver or passengers from the battery even in a serious accident.

A high temperature battery must be capable of many freeze/thaw cycles, as these will undoubtedly occur during the life of the battery from a variety of causes. At the present time all cells undergo four such cycles before they become operational, and several batteries have been cooled down and re-heated in operation without any cell failures occurring. These tests have

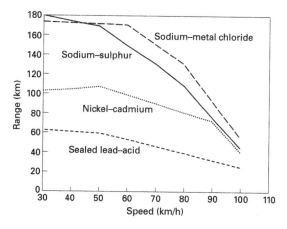

Fig. 8.28 Performance of converted Mercedes Benz 190 cars

Fig. 8.29 Battery drop test

been carried out at different states of charge and this does not appear to be a critical factor.

Technical challenges

Technical challenges are twofold: to increase the specific power of the battery towards the end of the discharge and to demonstrate in the pilot phase that the processes which have been developed for cell and battery production are capable of meeting the cost targets.

The increase in internal resistance which occurs during discharge is the result of the way in which the positive electrode operates – a reaction front moves through the electrode, starting at the β-alumina interface on both charge and discharge. This results in the ionic path through the melt getting longer and the ohmic resistance increasing. There are several ways of changing the cell geometry to reduce the electrode thickness. One possibility is to reverse the cell configuration, putting the positive electrode outside the β-alumina tube. This is something which has been considered. However, even if this was a satisfactory technical solution it would mean re-establishing the life and reliability data which have already been demonstrated for the central positive cell design. Other changes to cell design which retain the central positive configuration are being evaluated and show promise of being able to increase specific power by a factor of 2 at 80% depth of discharge. The challenge is to adapt the cell production processes in a way which is cost-effective.

9 Miscellaneous cells

Bruno Scrosati

9.1 Solid state cells

Introduction

The concept of an all-solid-state battery is very appealing since such a system should possess a number of very desirable characteristics, e.g. absence of any possible liquid leakage or gassing, the likelihood of extremely long shelf life and the possibility of operation over a wide temperature range. Solid state batteries could be constructed with excellent packaging efficiency for the active components, without separators and using simple lightweight containers. The opportunities for extreme miniaturization and very simple fabrication techniques are of obvious importance in applications where size and reliability are key factors, as for example in implantable electronic instrumentation such as cardiac pacemakers, physiological monitoring/telemetry packages, etc. However, to be of practical value, a solid state cell must also fulfil the basic requirements of all effective power sources, namely sufficiently low internal resistance and high voltage to give adequate levels of power, together with an acceptable cost. The first of these requirements is primarily a function of the electrolytic phase, which must therefore be a solid material characterized by high ionic conductivity with negligible electronic mobility; the former leads to low internal resistance of the cell, and the latter prevents self-discharge, thus promoting long shelf life.

There are two classes of materials which may be used as electrolytes in all-solid-state cells: polymer electrolytes, materials in which metal salts are dissolved in high molar mass coordinating macromolecules or are incorporated in a polymer gel, and ceramic crystalline or vitreous phases which have an electrical conductance wholly due to ionic motion within a lattice structure. The former were described in Chapter 7; in this chapter attention will be focused on the second class of solid electrolyte. These latter phases have been known for over a century, but have generally had high resistivity at ambient temperatures. An important development occurred towards the end of the 1960s with the discovery of a series of solids of general formula MAg_4I_5 (for M = Rb, K, . . .), which had exceptionally high ionic conductivity at room temperature (> 10 S/m).

A number of structural features were found to characterize these solids, which distinguished them from the more usual ionic crystals. Ionic transport in 'normal' salt lattices at high temperatures takes place by a defect mechanism, and the relatively poor conductivity of such phases is largely due to the small equilibrium concentration of defects present. Generally the structures of ceramic solid electrolytes are not close-packed, but contain two- or three-dimensional networks of passageways interlaced between interconnected polyhedra of the fixed ions, and through which selected mobile ions may move. The number of sites available for the mobile ions is much larger than the number of mobile ions themselves, and the solid therefore has a characteristic highly disordered structure. In $RbAg_4I_5$, for example, the 16 silver ions in each unit cell are distributed over 56 available sites.

The high conductivity is due to a combination of:

- a high concentration of mobile ions;

- a low activation energy for ionic motion from site to site.

One of the main advances in understanding such solid electrolytes stemmed from crystallographic studies which identified the types and numbers of sites available to the diffusing ions and emphasized the significance of the spatial relationship between sites. Mobility is enhanced when the mobile ion sites form a network of channels or a plane through which the ions can move, and the 'simplicity' of these passageways has an important influence on the value of the conductivity. In the crystalline solid electrolyte $Ag_{26}I_{18}W_4O_{16}$, the conduction pathways involve 90 face-sharing iodide polyhedra and 56 mixed iodide–oxide polyhedra per unit cell. Some silver ion pathways in this structure are shown in Fig. 9.1. The mobile ions are considered to move between low energy, highly coordinated sites by passing through a shared polyhedral face of only slightly higher energy. In Fig. 9.2, a number of interlinked tetrahedra formed by fixed ions are shown. A mobile ion traversing this structure in a vertical direction would pass through trigonal coordination on its way between neighbouring tetrahedral sites.

The second necessary condition for crystalline or vitreous solid to have high ionic conductivity is that the mobile ions have a high diffusion coefficient, i.e. it is indeed a 'fast ion conductor'. Much attention has been given to developing models of ionic motion. The simple hopping models applied successfully in the case of defect transport are not appropriate because of the high density of mobile ions in solid electrolytes, and

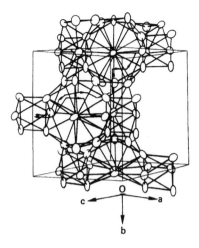

Fig. 9.1 The iodide arrangement in crystalline $Ag_{26}I_{18}W_4O_{16}$: some pathways for silver ion motion are shown by the arrows. (By permission of Academic Press: L.Y.Y. Chan and S. Geller, *J. Solid State Chem.*, 1977, **21**, 331.)

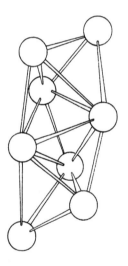

Fig. 9.2 Interconnected iodide ion tetrahedra providing a passageway for ion migration: mobile silver ions must pass through a three-coordinated configuration on their way between tetrahedral sites

the consequent correlation in their motion. Quasi-liquid sublattice models in which local oscillatory motion takes place during residence in favourable low energy locations and during transit between such sites have successfully explained experimental observations in a number of cases.

Since the discovery of the MAg_4I_5 compounds, very many more solid electrolytes have been reported. Sodium, lithium, copper, proton, oxide and fluoride solid electrolytes are now well known. An important example is sodium β-alumina which was discussed in Chapter 8. The conductance

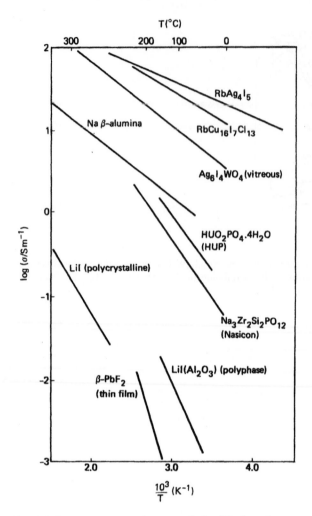

Fig. 9.3 The conductance of some typical solid electrolytes as a function of temperature

values of selected examples of solid electrolytes, of various types, are compared as functions of temperature in Fig. 9.3.

However, high electrolyte conductivity on its own does not necessarily guarantee low polarization in a solid state cell. Electrode/electrolyte interfacial resistance must also be taken into account, and in contrast to the more familiar situation with conventional aqueous systems where the solid electrodes are uniformly wetted by the liquid electrolyte, the all-solid configuration of the cell may create non-uniform contact at the interfaces. Differential expansion and contraction of electrodes and electrolyte may lead to poor contact (and consequent high internal resistance due to low effective electrode/electrolyte interfacial area) or even to a complete open circuit during cell operation. The situation is even more serious with secondary cells, as illustrated schematically in Fig. 9.4, where the effects

Discharge $M \rightarrow M^{n+} + ne$

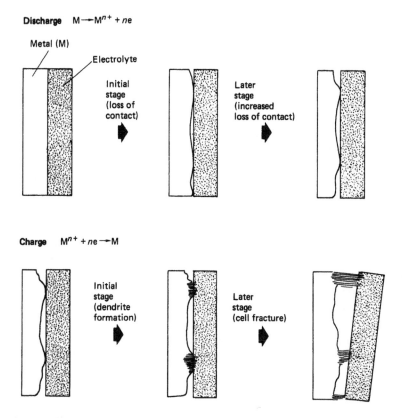

Fig. 9.4 Schematic diagram showing the effect of charge and discharge on a metal (M)|solid electrolyte (E) interface

of non-uniform discharge and the deposition of metal in dendritic form during charge are shown. Indeed, interfacial polarization phenomena are among the most severe problems in the development of practical solid state cells. A common expedient is to mix the electrode reactant with the electrolyte in order to form an electrodic mass (e.g. as a compressed powder) with a greatly enlarged interfacial area. This reduces the current density at the interface during operation of the cell and thus alleviates the polarization.

A wide variety of solid electrolytes with mobile silver ions have been reported, many of them having high conductivity and good stability. The majority of these electrolytes are based on silver iodide and may be regarded as 'cation modified', e.g. $RbAg_4I_5$, or 'anion modified', e.g. $Ag_7I_4AsO_4$. The decomposition potential of these materials (i.e. the voltage difference which produces significant electrolysis) is found to be very close to that of silver iodide itself, viz. 0.69 V at 25°C. This restricts the usable emf of single silver solid state cells to a value which is generally too low to meet the requirements of most electronic instrumentation. Commercial exploitation of silver-based cells is also inhibited by the high cost of silver.

Cells based on the transport of ions which have higher energy and lower cost have also been studied. A number of copper solid electrolytes have been characterized, but their stability has generally been found to be poor, especially in the presence of air or moisture. More attention has been paid to the search for lithium solid electrolytes because of the possibility of developing a high voltage solid state cell which might have a performance comparable with the lithium–organic cells described in Chapter 4. However, despite reports on a wide range of inorganic crystalline or vitreous lithium solid electrolytes, no really good ambient lithium ion conductor has yet been discovered.

In its most general configuration, a solid state cell may be formed by finely grinding the constituents and then sequentially pressing the anode–electrolyte mixture, the electrolyte and the cathode–electrolyte mixture (with or without 5–10% of a binding agent) in a suitable die to form a single composite pellet. When the electrolyte is of the ceramic type this procedure is not feasible and a sintering process has to be used instead. Since the electrolyte does not participate in the cell reaction it is sometimes possible to reduce its thickness to that of a thin film. Under these circumstances it is possible to replace a highly conductive solid electrolyte with, say, a defect ionic conductor of much higher resistivity.

The only commercial ambient solid state batteries so far produced have been based on either silver or lithium anodes and these will now be described.

Silver–iodine cells

The first commercial solid state battery was manufactured at the end of the 1960s in the USA by Gould Ionics. This was a silver–iodine battery using $RbAg_4I_5$ as electrolyte. An essential constraint on any cell system is that the active components must not react with the electrolyte either directly or by electrolytic action. Free elemental iodine reacts with $RbAg_4I_5$, degrading it to poorly conducting phases by the process

$$I_2(s) + RbAg_4I_5(s) \rightarrow RbI_3(s) + 4AgI(s) \qquad (9.1)$$

Thus the cell

$$Ag(s)|RbAg_4I_5(s)|I_2(s),C(s)$$

cannot be formed in an equilibrium condition. Further, since the emf of such a cell at 25°C would be 0.687 V, which is higher than the decomposition potential of $RbAg_4I_5$ (0.660 V), an electrolytic process such as

$$RbAg_4I_5(s) \rightarrow 2Ag(s) + 2AgI(s) + RbI_3(s) \qquad (9.2)$$

would occur in the cell. The iodine must therefore be complexed to reduce its activity. In the Gould system, rubidium triiodide was used as the iodine source, to give the cell

$$Ag(s)|RbAg_4I_5(s)|RbI_3(s),C(s)$$

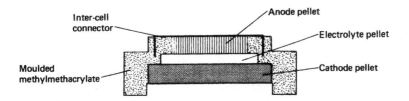

Fig. 9.5 Ag–RbI$_3$ solid state cell. (By permission of John Wiley: B.B. Owens, *Advances in Electrochemistry and Electrochemical Engineering*, vol. 8, eds P. Delahay and C.W. Tobias, 1971, p. 1.)

which has an emf of 0.660 V at 25°C and a theoretical energy density of 48 Wh/kg. Since RbAg$_4$I$_5$ is thermodynamically unstable at temperatures below 27°C, two overall cell reactions have been proposed:

Above 27°C:

$$14Ag(s) + 7RbI_3(s) \rightarrow 3RbAg_4I_5(s) + 2Rb_2AgI_3(s) \qquad (9.3)$$

Below 27°C:

$$4Ag(s) + 2RbI_3(s) \rightarrow Rb_2AgI_3(s) + 3AgI(s) \qquad (9.4)$$

A typical configuration of the Gould silver–iodine battery is shown in Fig. 9.5. The anode is a blended mixture of powdered silver, carbon and electrolyte. High efficiency coupled with low polarization is obtained by forming the electrode by *in situ* reduction of Ag$_2$O by carbon. The cathode is a blended mixture of RbI$_3$, RbAg$_4$I$_5$ and carbon. The electrolyte layer may contain up to 10% of a polycarbonate resin as binding agent. The cell components are sequentially pressed into a single three layer pellet. A series of pellets can be combined to form a higher voltage battery, as shown in Fig. 9.6. Because the electrolyte disproportionation reaction

$$2RbAg_4I_5(s) \rightarrow 7AgI(s) + Rb_2AgI_3(s) \qquad (9.5)$$

is catalysed by water vapour, the cell components must be handled in a dry atmosphere and the battery casing must be hermetically sealed.

One of the main features of these silver–iodine batteries is their extremely long shelf life. Cells stored for more than 20 years have been shown to deliver practically their full rated capacity. On the other hand, these batteries had a very limited range of application because of their low voltage, modest energy value and high cost, so that their production soon ceased. However, the Gould silver–iodine battery is of great historical importance as it greatly stimulated the development of solid state systems.

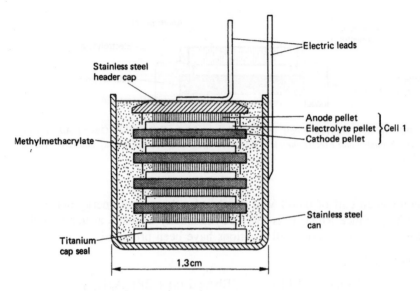

Fig. 9.6 Ag–RbI$_3$ solid state battery. (By permission of John Wiley: as Fig. 9.5.)

Lithium–halogen cells

The solid state battery which has had the greatest commercial success is the primary lithium–iodine cell, developed originally by Catalyst Research Corporation in the USA and now also manufactured by Wilson Greatbatch Ltd., and Medtronic Inc. In this cell, the lithium iodide electrolyte has a relatively high resistance (10^5 Ωm at room temperature) and should perhaps not be considered as a 'true' solid electrolyte since electric transport occurs by a lithium ion vacancy diffusion mechanism. The anode is solid lithium metal and the cathode is an iodine charge transfer complex formed by combination of a poly-2-vinylpyridine donor (here abbreviated to P2VP) and iodine as acceptor. An interesting property of this type of complex is that the electronic conductivity is many orders of magnitude higher than that of either donor or acceptor, and that this high conductivity is maintained over a wide range of donor/acceptor compositions. The complex is easily formed by direct reaction of iodine with P2VP powder to form a highly viscous tar which can be poured when hot, or a semi-solid mass. It generally contains over 90% by weight of iodine and has an iodine vapour pressure close to that of elemental iodine, so that it makes a very convenient iodine electrode.

Lithium–iodine cells are produced simply by making direct contact between anode and cathode. On touching, a thin layer of lithium iodide is formed by direct reaction. As soon as the layer becomes complete, the reaction rate decreases sharply, since diffusion of iodine through lithium iodide is very slow. Thus the cell may be written as

$$Li(s)|LiI(s)|P2VP.nI_2(s)$$

That the cell reaction may be written simply as

$$Li(s) + \tfrac{1}{2}I_2(s) \rightarrow LiI(s) \qquad (9.6)$$

is supported by its measured emf of 2.80 V at 25°C, since this value agrees exactly with that calculated for the reaction. Latimer's *Oxidation potentials* gives the free energy of formation of LiI(s) at 25°C as 268 kJ/mol, corresponding to an emf of 2.79 V. More accurately, the cell reaction might be written as

$$14Li(s) + P2VP.8I_2(s) \rightarrow 14LiI(s) + P2VP.I_2 \qquad (9.7)$$

On discharge, lithium ionizes at the anode/electrolyte interface and lithium ion transport through the lithium iodide occurs by the motion of a lithium ion vacancy in the direction from cathode to anode, as shown schematically in Fig. 9.7. As the discharge progresses, further LiI forms at the cathodic side of the electrolyte, and as the LiI layer thickens, so the internal resistance of the cell rises, as shown in Fig. 9.8. The resistivity of the P2VP.nI$_2$ complex also varies during the discharge, since it depends on iodine content. A marked increase in resistance occurs when the I$_2$/P2VP ratio reaches about 2, after which there is an abrupt decay in cell voltage, which indicates the end of the useful operating range of the cell.

Self-discharge by the diffusion of iodine through the LiI layer is minimal. The result of this is that only a very thin layer of LiI is initially formed, so that the resistance of a freshly made cell is only 2–3 kΩ, and losses in capacity during storage or use are small – estimated to be 5–10% over a period of 10 years. On the other hand, if the lithium electrode should

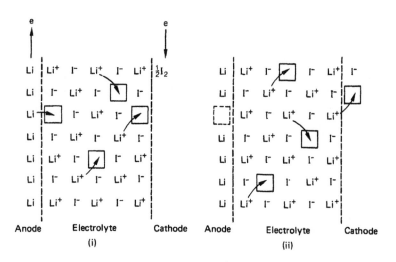

Fig. 9.7 Mechanism of lithium ion transport by vacancy motion in lithium iodide. Cell discharge is accompanied by the migration of lithium ion vacancies in the direction from cathode to anode. (By permission of John Wiley: as Fig. 9.5)

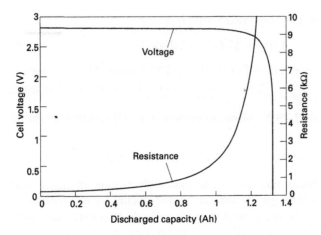

Fig. 9.8 Variation of cell voltage and internal resistance during discharge of a Li(s)|LiI(s)|P2VP.nI_2(s) cell. (By permission of Catalyst Research Corporation.)

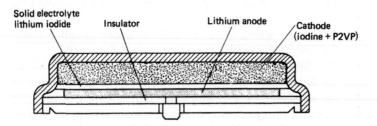

Fig. 9.9 Lithium–iodine solid state button cell for electric watches and pocket calculators. (By courtesy of Catalyst Research Corporation.)

become internally short-circuited to the cathode, a fast chemical reaction would quickly remove the short. The cells are also self-protected against external short circuits since, by their nature, they are current-limited. Cell polarization during discharge can be significantly reduced by coating the lithium surface with a film of P2VP. It is now clear that this improvement is due to a fundamental modification of the LiI layer which no longer acts as a 'true' solid electrolyte.

Lithium–iodine cells are assembled in a controlled environment and are hermetically sealed with welded cases and glass-to-metal or ceramic-to-metal seals. They are manufactured in three basic configurations: as normal button cells, as button cells for direct mounting on printed circuit boards, and as cardiac pacemaker batteries. In Fig. 9.9 the cross-section of a button cell, used mainly in electric watches and pocket calculators, is shown. Such cells are produced with rated capacities in the range 120–250 mAh and are designed for current drains of 0–60 μA. At low drains (1μA/cm^2), a practical energy density of 120 Wh/kg (600 Wh/dm^3) is obtained. The cells may be operated over a wide temperature range, although the max-

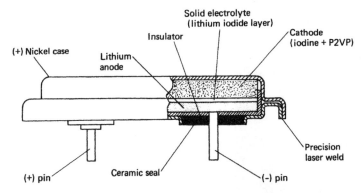

Fig. 9.10 Construction of lithium–iodine solid state cell for PCB mounting. (By courtesy of Catalyst Research Corporation.)

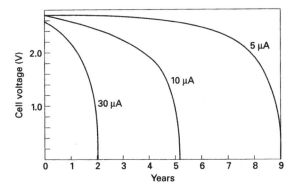

Fig. 9.11 Projected discharge characteristics of a nominal 400 mAh lithium–iodine solid state cell. (By permission of Catalyst Research Corporation.)

imum discharge current falls off logarithmically with decreasing temperature.

Larger button cells with rated capacities in the range 35–1000 mAh are manufactured for direct mounting on printed circuit boards where they are used as standby power sources for CMOS RAMs, reference voltage sources, etc. Fig. 9.10 shows the construction of these cells and the position of the terminal/mounting pins. Projected discharge curves for Catalyst Research Corporation cells are given in Fig. 9.11.

The high reliability and the complete absence of faults such as electrolyte leakage or gas generation make the lithium–iodine solid state battery a particularly suitable device for powering implanted electronic devices, and it is now widely used in the cardiac pacemaker industry. The design and construction of Medtronic Inc. pacemaker batteries are shown in Fig. 9.12. A typical unit, such as the 'Enertec Alpha 33', has dimensions of 33.4 mm × 27.4 mm × 7.9 mm, giving a total volume of 6.0 cm^3 and a mass of 22 g. The cell has a completely welded construction and uses a specialized

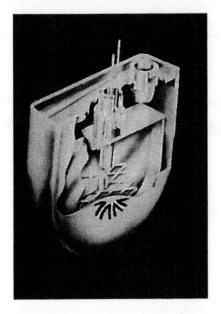

Fig. 9.12 Cutaway of a lithium–iodine pacemaker cell. (By courtesy of Medtronic Inc.)

glass-to-metal seal for the electrical feedthrough. The projected capacity as a function of current drain, determined on the basis of a mathematical model, is shown in Fig. 9.13. The upper line represents the theoretical capacity of 2.6 Ah, assuming 100% utilization of the iodine present in the P2VP.8I_2; since each vinyl pyridine entity reacts irreversibly with approximately 0.5 mol of iodine, the maximum available capacity is reduced to 2.1 Ah, giving a maximum energy density of 290 Wh/kg (960 Wh/dm^3). The practical capacity depends on the rate of discharge. At very low rates (10 μA) self-discharge processes limit the capacity, while at high rates (100 μA) polarization becomes dominant.

Other lithium-based solid state cells

Mixtures of ion-conducting and non-conducting phases have been termed polyphase solid electrolytes. Mixing LiI with approximately 50 m/o γ-Al$_2$O$_3$ increases the conductivity of the lithium ions by a factor of about 20 according to some workers, and of over 100 according to others. The origins of this enhancement in conductivity are not clear and cannot be explained by any classical doping mechanism which would require an increase in the number of lithium ion vacancies in the lattice. Despite the complete stability of the polyphase system to lithium, the presence of small amounts of water in the alumina particles has been suggested as an important factor. It is considered that adsorbed water might react with the lithium iodide to form a highly conductive monohydrate phase. It is known that the conductance of LiI increases by about two orders of magnitude

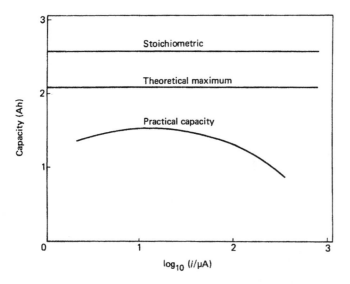

Fig. 9.13 Projected practical capacity of a lithium–iodine pacemaker battery (Enertec Alpha 33) as a function of current drain. (By courtesy of Medtronic Inc.)

when passing from the anhydrous salt to $LiI.H_2O$. However, it is unlikely that the contribution of trace levels of water alone can explain the enhancement in conductivity observed in $LiI(Al_2O_3)$ and it may be that a mechanism involving increased grain boundary or surface conductivity, favoured by the high surface area of the dispersed alumina, should be considered.

The enhancement in conductivity obtained by dispersing Al_2O_3 in the lithium iodide has permitted this composite electrolyte to be used in the fabrication of cells in the form of compressed pellets without the creation of prohibitive values of internal resistance at ambient temperatures. An example is given in Fig. 9.14 which shows a schematic cut-away view of the practical cell developed by P.R. Mallory & Co (now Duracell International) in the 1970s.

The cell used a lithium anode and a lead iodide–lead sulphide cathode:

$$Li(s)|LiI(Al_2O_3)(s)|PbI_2(s),PbS_2(s),Pb(s)$$

The $LiI(Al_2O_3)$ electrolyte layer was approximately 0.2 mm thick and the cell pellet was retained in a polypropylene ring, and sealed in a steel case. The cell had an OCV of 2.0 V at 25°C, dropping to 1.91 V after the first few per cent of the discharge. The discharge reactions are

$$2Li(s) + PbI_2(s) \rightarrow 2LiI(s) + Pb(s) \tag{9.8}$$

and

$$2Li(s) + PbS(s) \rightarrow Li_2S(s) + Pb(s) \tag{9.9}$$

Because of its long-term stability and reliability, this cell found some application in the cardiac pacemaker field.

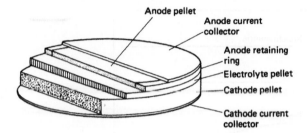

Fig. 9.14 Cross-section of a Li(s)|LiI(Al$_2$O$_3$)(s)|PbI$_2$(s),Pb(s) solid state cell. (By permission of the Electrochemical Society: C.C. Liang and L.H. Barnette, *J. Electrochem Soc.*, 1976, **123**, 453.)

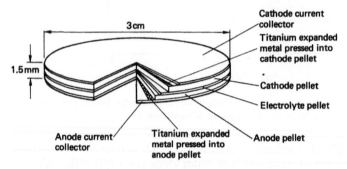

Fig. 9.15 Cross-section of the Li$_n$Si(s)|LiI(Al$_2$O$_3$)(s)|TiS$_2$(s),SbS$_3$(s),Bi(s) solid state cell

Some successful development of rechargeable solid state systems was achieved by using lithium intercalation cathodes, such as TiS$_2$, which operate in exactly the same manner as in the lithium–organic cells described in Chapter 7. One example of this type of cell is provided by the battery system developed in the 1970s by P. R. Mallory and Co. (now Duracell) based on the following scheme:

$$\text{Li}_n\text{Si(s)|LiI(Al}_2\text{O}_3\text{)(s)|TiS}_2\text{(s)LiI(Al}_2\text{O}_3\text{)(s),C(s)}$$

As illustrated schematically in Fig. 9.15, the cell was assembled by pressing the components into a single pellet, which was then hermetically sealed in a stainless steel case with a glass-to-metal seal. The advantage of a 'soft' electrolyte such as LiI(Al$_2$O$_3$) is that an effective cell could be made without the need for high temperature sintering processes, as good inter-particle contact was achieved by cold pressing. In addition, some volume changes could be accommodated.

The cell operated at 300°C, at which temperature lithium metal is liquid, so that the lithium anode was replaced by a lithium–silicon alloy which is solid and which exhibits a reversible uptake of lithium, as discussed in Chapter 8.

Optimum behaviour was obtained with a cathode mixture comprising the intercalation host with electrolyte and graphite which, respectively, reduced contact polarization and enhanced the electronic conductance.

The emf of this cell was 2.4 V at 300°C and the cell reaction is given by

$$Li_nSi(s) + xTiS_2(s) \rightleftharpoons Li_{n-1}Si(s) + Li(TiS_2)_x(s) \qquad (9.10)$$

Over 300 cycles at current densities of 5 mA/cm^2 were obtained, with efficiencies of 80–90% for a 400 mAh rated cell. The practical energy density was 280 Wh/kg.

The good cycling behaviour and energetic characteristics, coupled with the inherent advantages of an all-solid-state configuration, made this system a promising candidate for large-scale storage applications. Accordingly, design studies were made for a 72 kWh prototype using 24 battery modules in parallel, each module containing 36 cells in series to give a 72 V system. The total energy density of the battery, including provision for heating elements and insulation, was estimated as 170 Wh/kg (310 Wh/dm^3). However, a lack of durability on prolonged cycling prevented this system becoming competitive with other high temperature storage batteries (Chapter 8).

Another type of lithium solid state cell which reached the quasi-commercial stage was developed in the mid-1980s by Eveready. This cell used a vitreous inorganic electrolyte formed by a mixture of LiI and $Li_8P_4O_{0.25}S_{13.75}$. This solid electrolyte had a reasonably high conductivity which allowed cell operation at ambient temperatures. A disadvantage was its high reactivity which imposed the use of severe fabrication controls (e.g. assembly in a strictly moisture-free environment).

The Eveready cell used a basic Li–TiS$_2$ couple in the following scheme:

$$Li(s)|solid\ electrolyte|TiS_2(s),solid\ electrolyte,C(s)$$

The emf of this cell was 2.4–2.6 V at room temperature, corresponding to a cell process which may be written as

$$xLi(s) + TiS_2(s) \underset{charge}{\overset{discharge}{\rightleftharpoons}} Li_xTiS_2 \qquad (9.11)$$

The battery is constructed in the button-type configuration and typical discharge characteristics are shown in Fig. 9.16. The rated capacity of this particular cell was 36 mAh and its reported energy density values were 44 Wh/kg and 150 Wh/dm^3.

Other developments in the area of solid state lithium batteries include prototype production and testing of thin-film microbatteries at Oak Ridge National Laboratory in the USA. The fabrication involves electrode and electrolyte film deposition to form compact layers of thickness of the order of few microns. The cell uses a lithium anode, an amorphous $Li_{3.3}PO_{3.9}N_{0.17}$ solid electrolyte and an amorphous V$_2$O$_5$ cathode:

$$Li(s)|Li_{3.3}PO_{3.9}N_{0.17}(s)|V_2O_5(s)$$

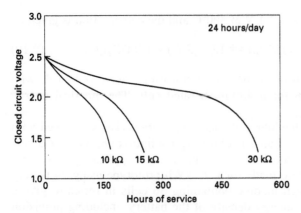

Fig. 9.16 Discharge characteristics at 21°C of the Li–TiS$_2$ solid state cell, type XR 2016. (By permission of Eveready.)

The cell emf is 3.6 V corresponding to

$$x\text{Li(s)} + \text{V}_2\text{O}_5\text{(s)} \underset{\text{charge}}{\overset{\text{discharge}}{\rightleftharpoons}} \text{Li}_x\text{V}_2\text{O}_5\text{(s)} \qquad (9.12)$$

A prototype battery of about 8 mm thickness and a total area of 1 cm^2 had a rated capacity of 150 mAh over the full voltage range, which extends from 3.6 to 1.5 V. The battery design is suitable for circuit integration as a standby power source for low current electronic devices.

9.2   Secondary hybrid cells

Introduction

A hybrid cell is defined here as a galvanic electrochemical generator in which one of the active reagents is in the gas phase. Hybrid cells occupy an intermediate position between the closed galvanic cells described in the remainder of this book, where operation is confined to reactants added to the cell at manufacture, and fuel cells in which both cathodic and anodic reactants are supplied continuously (usually in gaseous form) from sources external to the cell. Hybrid cells take advantage of both battery and fuel cell technologies. The most common example of a primary hybrid cell, namely the metal–air system, was considered in Chapter 3. In 'mechanically recharge-able' cells, the spent metal anode is substituted by a new electrode at the end of the discharge. Here, some typical rechargeable hybrid cells are described.

Metal–air cells

Metal–air cells have a very favourable energy density which is achieved through not requiring to incorporate the positive active component within

Table 9.1 Emf and theoretical energy density of practical metal–oxygen cells

Couple	Reaction	Emf (V)	Theoretical energy density (Wh/kg)
Zn/O_2	$Zn(s) + \frac{1}{2}O(g) \rightarrow ZnO(s)$	1.65	1090
Fe/O_2	$Fe(s) + \frac{1}{2}O_2(g) \rightarrow FeO(s)$	1.27	970

the cell. To a large extent, the impetus for research and development in this field has arisen from possible EV applications where energy density is a critical parameter. Metals considered as possible secondary negative electrodes are confined to zinc and iron; others have been excluded on the grounds of cost or weight. The emf and theoretical energy density of zinc–oxygen and iron–oxygen couples are given in Table 9.1. Commercial interest in secondary zinc–air cells is very high.

The oxygen electrode

The principal features of the electrochemistry and construction of oxygen electrodes in alkaline solution, which were considered in Chapter 3 for primary cells, are common to secondary cells. If air is used rather than oxygen, it is then necessary to scrub the gas to remove carbon dioxide, since otherwise the electrolyte becomes progressively contaminated with carbonate which reduces the conductivity and may block electrode pores.

The main problem associated with oxygen electrode development for secondary batteries is related to the behaviour of the electrocatalyst during recharge, when it is necessary to apply high anodic voltages. Inexpensive catalysts based on carbon or silver cannot be used as they are rapidly oxidized. Even noble metal catalysts show significant dissolution or deep surface oxidation on recharge, which reduces their catalytic efficiency for the discharge reaction and hence increases the oxygen electrode polarization. One solution to this problem is the use of a third (auxiliary) electrode of perforated metal or wire mesh which carries all of the anodic current during charging. This procedure requires the mechanical switching of every cell in a battery twice a cycle and is completely unsuitable for high power batteries since the size and cost of the switches would be prohibitive. An alternative method developed by Lucas, Siemens, Swedish National Development Co., and others is to use a composite electrode structure (Figs 9.17a and b). Here the auxiliary electrode is in parallel with their air electrode. During discharge the auxiliary is inactive, but during charging a pocket of oxygen bubbles is formed which effectively isolates the active oxygen electrode from the electrolyte. In Fig. 9.17(a) the fine wire mesh on the solution side of the electrode fulfils the same function as the coarse pored structure in Fig. 9.17(b).

The oxygen electrode suffers from considerable polarization losses on discharge, largely due to mass transport limitations. Metal–air cells have

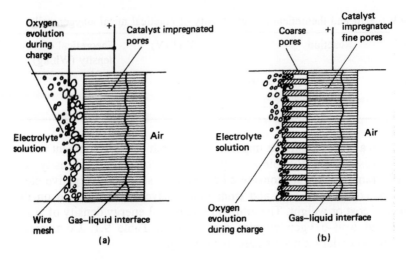

Fig. 9.17 Methods for using layers of oxygen bubbles to isolate active oxygen electrode surfaces from electrolyte during charging

consequently poor high rate performance. Polarization is also high on charge, so that cycle energy efficiency is very low. Further problems associated with this electrode system include water loss, especially at elevated temperatures, and leakage of electrolyte.

Iron electrodes and iron–air cells

The iron–oxygen cells may be represented as

$$Fe(s)|KOH(aq)|O_2(g),M(s)$$

where M is the electrocatalyst, and the principal cell reactions are

$$Fe(s) + H_2O(1) + \tfrac{1}{2}O_2(g) \underset{charge}{\overset{discharge}{\rightleftharpoons}} Fe(OH)_2(s) \qquad (9.13)$$

Further electrochemical oxidation to FeO(OH) may also take place. Dendrites are not formed on charging since the solubility of $Fe(OH)_2$ is low, but considerable hydrogen evolution takes place which lowers the cycle efficiency. The iron electrode also suffers from very high self-discharge ($\approx 2\%$ per day at 25°C), due to the reactions

$$H_2O(1) + e \rightarrow \tfrac{1}{2}H_2(g) + OH^-(aq) \qquad (9.14)$$

and

$$Fe(s) + 2OH^-(aq) - 2e \rightarrow Fe(OH)_2(s) \qquad (9.15)$$

Some reduction in corrosion and improvement in cycle life has been brought about by additives such as sulphide ion. The best electrodes have a very similar configuration to the sintered plate electrodes used in iron–nickel oxide cells, and described in Chapter 6.

Iron–air cells have been developed by Matsushita Battery Industrial Co. and by the Swedish National Development Co., which have given an energy density of 80 Wh/kg at C/5 and a cycle life of 200 cycles to 60% depth of discharge. The latter company have produced 15–30 kWh batteries for EV testing. One limitation of the iron–air system for this application is the low power density achieved – a maximum value of 30–40 W/kg is reported. Similar cells are also being developed by Westinghouse (USA) and Siemens (Germany).

Zinc electrodes and zinc–air cells

The zinc–air cell,

$$Zn(s)|KOH(aq)|O_2(g),M(s)$$

has a higher energy density than its iron equivalent, but to the problems of the air electrode are added those of recharging the zinc electrode (Chapter 6). The cell reactions are formally

$$Zn(s) + H_2O(1) + \tfrac{1}{2}O_2(g) \underset{\text{charge}}{\overset{\text{discharge}}{\rightleftharpoons}} Zn(OH)_2(s) \qquad (9.16)$$

but the main oxidation product of zinc is probably the $Zn(OH)_4^{2-}$ ion. A number of techniques have been tested with a view to improving the quality of the zinc deposit, such as electrolyte circulation, electrode vibration and the use of air bubblers to disturb the diffusion layer. Energy densities of over 130 Wh/kg at C/5 have been claimed, but it is generally accepted that the zinc electrodes deteriorate fairly rapidly with cycling. Crucial tasks for the successful production of rechargeable zinc–air batteries of practical importance include the development of kinetically fast and durable oxygen electrodes and the identification of the best technical approaches for ensuring long cyclability of the zinc electrode. Recent achievements in these aspects have been the use of bifunctional catalysts for oxygen evolution and of a moderately alkaline ZnO-saturated electrolyte for smooth zinc deposition.

Many battery companies have attempted the production of zinc–air battery prototypes. Encouraging results were obtained by CGE in France with the system shown schematically in Fig. 9.18.

A slurry of zinc powder in aqueous KOH was pumped through the cyclindrical cell whose outer surface comprised the air electrode and for which the negative current collector was an axial brass rod. The slurry was recirculated through the cell several times until an acceptable percentage of the theoretical capacity had been utilized (i.e. for zinc oxidation to soluble zincate). For recharge, the zincate ion was reduced to zinc powder either externally or in an electrolysis cell connected to the system. In this context

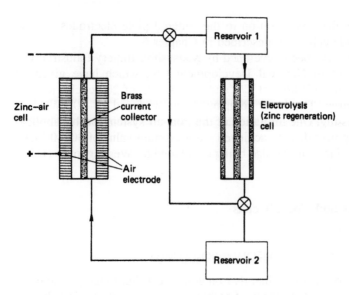

Fig. 9.18 Schematic diagram of a zinc–air secondary battery system

it was an advantage to deposit the zinc in dendritic form: the slurry could be reconstituted by momentarily reversing the current to detach the dendrites. The complete CGE system, including recharging cell and pumps, was reported to have a practical energy density of over 90 Wh/kg. Cell lifetimes of over 1000 cycles were obtained, with batteries of 12 series connected cells giving 500–600 cycles. However, the complexity of the system together with the low-specific power is likely to limit its exploitation as an EV power source.

A more recent attempt is in progress at AER Energy Resources Inc. in the USA. This company is developing a zinc–air prototype under the tradename of AER Energy™. The battery uses an innovative technology which contains an air breathing, bi-functional cathode, a zinc anode and a potassium hydroxide electrolyte. The AER Energy battery system, which has been reported to have an energy density of 98 Wh/kg at a C/8 rate, is designed to provide power for portable computers and similar consumer electronic devices.

Zinc–air modules for EV application are under development at the Edison company in Italy and by the Electric Fuel Ltd in Israel. In this case, the battery recharge also includes a mechanical step, namely the removal and replacement of the spent zinc electrodes. The actual electrochemical recharging process is carried out in a remote station. The proposed application to passenger vehicles considers the construction of specific stations where the removal and replacement of the spent zinc electrode pack is carried out automatically (Fig. 9.19). Energy and power densities of the order of 200 Wh/kg and 100 W/kg, respectively, and long cyclability, which may provide the car with a 300 km range and a

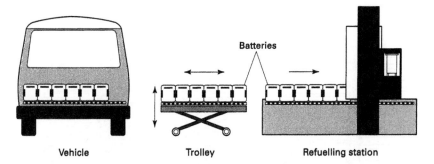

Fig. 9.19 Schematic diagram of a 'filling station' for electric vehicles powered by mechanically rechargeable zinc–air batteries. (By permission of Edison.)

100 000 km life, have been predicted and confirmed in prototype demonstration vehicles.

Metal–halogen cells

The concept of an aqueous zinc–chlorine cell system brings together a number of attractive features. First, chlorine, as well as being abundant and inexpensive, has a highly positive equilibrium reduction potential, so that cells with chlorine anodes have a high energy density. Further, chlorine is fairly soluble in aqueous solution (approximately 0.1 mol/dm^3 at ambient temperatures, but dependent on the other solution components), so that three-phase fuel cell electrode technology is not required. From the standpoint of the zinc electrode, zinc deposition morphology is much more favourable in acid media, with few dendrite problems unless at very high current densities. The cell is represented as

$$Zn(s)|ZnCl_2(aq)|Cl_2(aq),M(s)$$

and the basic cell reactions are

$$Zn(s) + Cl_2(g) \underset{charge}{\overset{discharge}{\rightleftharpoons}} ZnCl_2(aq) \tag{9.17}$$

The standard emf is 2.12 V at 25°C.

The key problem associated with this system is the storage of chlorine which is highly corrosive and difficult to manipulate in either gaseous or liquid form. In the batteries which were developed for EV and load levelling applications by Energy Development Associates (Gulf + Western Industries in conjunction with Occidental Petroleum Corporation) in the USA, this difficulty was overcome by converting the chlorine to the solid hydrate, $Cl_2.6H_2O$, which is stable below 9.6°C. This material is not particularly corrosive and may be readily stored in a refrigerated chamber. By controlling the temperature of this store, the system may be operated

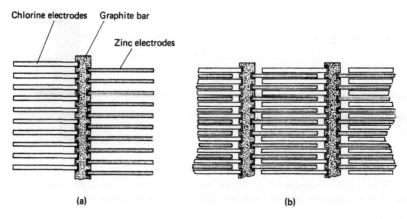

Chlorine electrodes   Graphite bar

Zinc electrodes

(a)   (b)

Fig. 9.20 Comb-type bipolar electrodes for zinc–chlorine batteries: (a) bipolar stack; (b) unit cell

close to atmospheric pressure, without the need for corrosion-resistant gas compressors.

The zinc electrode does not present the problems associated with charging in alkaline solutions. However, the nature of the deposit is affected by factors such as current density, flow conditions, electrolyte concentration, etc., and various additives are used to minimize the incidence of dendrite formation. Zinc corrosion accompanied by hydrogen evolution is a significant problem because of the possibility of explosive hydrogen/chlorine mixtures being formed. Hydrogen evolution rates depend on the impurities present in the system and the pH of the electrolyte, which in turn is affected by zinc ion hydrolysis reactions and hence by $ZnCl_2$ concentration and temperature. Accumulation of hydrogen is avoided by regulating the electrolyte composition and by initiating controlled recombination using UV radiation.

Development of chlorine electrode materials has benefited from the experience of chlor-alkali electrolysis cell technology. The main problem is to find the best compromise between cycle life and cost. Porous graphite, subjected to certain proprietary treatments, has been considered a preferable alternative to ruthenium-treated titanium substrates. The graphite electrode may undergo slow oxidative degradation, but this does not seem to be a significant process.

The design of practical batteries based on comb-type bipolar stacks is shown diagrammatically in Fig. 9.20(a). Machined graphite bus bars form the electrical connections between the electrodes on each stack and also separate the individual cell electrolytes. The zinc and chlorine electrode substrates are fitted into grooves on either side of the bus, and the battery is assembled by interleaving the comb structures as shown in Fig. 9.20(b). The stacks are terminated with monopolar units at each end. A channel is cut through the centre of the chlorine electrode substrate which enables electrolyte to be pumped through the porous mass during the charge or discharge process. The electrolyte then flows up between the zinc and chlorine electrodes and out at the end of each cell before returning to the pumping system. A great deal of work was carried out in order to optimize the dimensions of the battery structure with a view to designing a system with a favourable current distribution and maximum capacity.

The complexity of the system and the dangers of chlorine evolution in the event of an accident disabling the refrigeration unit has militated against its application in EVs. However, 10 MWh load levelling modules have been designed and are projected to have energy efficiencies of 70–75%, together with a very competitive capital cost per kW.

The zinc–bromine system has a much lower energy density than that of zinc–chlorine, but the comparative ease of handling bromine and the absence of refrigeration requirements make it an attractive alternative. The relatively low power and energy density, coupled with the need for pumps, reservoirs, etc., probably rule out EV applications, but numerous development programmes for load levelling batteries are being actively pursued by a number of companies, formerly by Exxon Research and Engineering Corp. and by Gould Inc. in the USA, and currently by Power-cell GmbH in Austria.

The cell is represented as

$$Zn(s)|ZnBr_2(aq)|Br_2(g),M(s)$$

and the overall cell reaction is written formally as

$$Zn(s) + Br_2(g) \underset{charge}{\overset{discharge}{\rightleftharpoons}} ZnBr_2(aq) \qquad (9.18)$$

The standard emf is 1.83 V at 25°C and the theoretical energy density is 430 Wh/kg. In practice, complex species such as Br_3^- are also involved, and the zinc ions also compete for the bromide ions to form a range of complexes from $ZnBr^+$ to $ZnBr_4^{2-}$. (At high zinc bromide concentrations the apparent cation transport number actually becomes negative.) KCl may be added to the electrolyte to maintain high conductivity as the $ZnBr_2$ concentration falls during charge. The bromine electrode substrate material M must maintain good reversibility for the electrochemical process, while at the same time resisting chemical attack by the bromine. A number of materials, such as titanium and ruthenium dioxide, perform very well but are expensive. Graphite cannot be used because of the formation of bromine intercalation compounds. Vitreous carbon resists attack and appears to be a very suitable material. While zinc can be deposited on a wide range of substrates, the choice in practice is very limited. As in the case of the zinc–chlorine cell, adequate control of zinc deposit morphology can only be retained by stripping off all the zinc from the positive electrode on all or most cycles. This requires that the cells are zinc-limited, and this in turn implies that in a battery, a number of cells must be reversed towards the end of a discharge to ensure complete stripping in the others. Under these

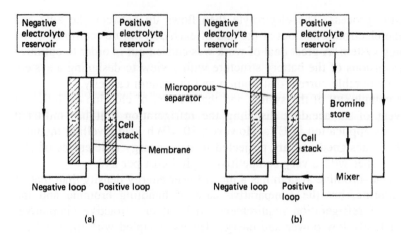

Fig. 9.21 Schematic diagrams of zinc–bromine battery systems: (a) cell with cation selective membrane; (b) cell with reservoir for polybromide and microporous separator

circumstances, bromine is evolved on the negative, and the substrate must be able to withstand this process. In practical batteries this problem is often solved by using bipolar vitreous carbon electrodes.

A particular problem with this cell system is the high rate of self-discharge due to the chemical reaction of bromine with the zinc negative. One solution is to separate anolyte and catholyte by a cation-selective membrane which is impermeable to bromine, such as perfluorosulphonic acid-based materials like 'Nafion', as illustrated in Fig. 9.21(a). An alternative approach is to react the bromine in order to form an insoluble solid or liquid phase. For example, unsymmetric quaternary alkyl ammonium perchlorates react with bromine to form oily polybromides which significantly reduce the level of free bromine in solution. In some designs the insoluble bromine complexes are retained within the porous electrode structure; in more advanced large capacity systems, a separate polybromide reservoir is incorporated (Fig. 9.21b). A simple microporous separator is then sufficient to prevent excessive self-discharge. A typical cell stack is shown schematically in Fig. 9.22. Practical cells now all employ electrolyte circulation to improve the quality of the zinc deposit and to permit the operation of heat exchangers. Note that in both cell configurations shown in Fig. 9.21(a) and (b), twin circulation systems are required. The Austrian company Powercell is currently developing zinc–bromine prototypes for EV use which have a projected energy density of 70–90 Wh/kg and a cycle life of more than 500.

Hydrogen–metal cells

As with the metal–oxygen system, hydrogen–metal cells can be considered as closed galvanic/fuel cell hybrids. They make use of the hydrogen

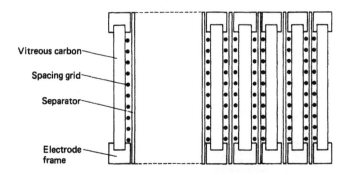

Vitreous carbon

Spacing grid

Separator

Electrode frame

Fig. 9.22 Cell stack for a zinc–bromine battery

electrode technology of fuel cells and that of nickel or silver electrodes from alkali secondary cells.

The charged hydrogen–nickel cell is represented as

$$Pt(s),H_2(g)|KOH(aq)|NiO(OH)(s)$$

and has an OCV of 1.5–1.6 V in its fully charged state. The associated cell reactions are

$$\tfrac{1}{2}H_2(g) + NiO(OH)(s) + H_2O(l) \xrightleftharpoons[\text{charge}]{\text{discharge}} Ni(OH)_2.H_2O(s) \qquad (9.19)$$

The cell or battery is enclosed in a stainless steel or Inconel pressure vessel (Fig. 9.23). Hydrogen pressure rises from about 0.5 MPa in the fully discharged state to 3–10 MPa when charged, and the pressure in the vessel can be used to monitor the state of charge. Direct reaction between hydrogen and nickel oxide is relatively slow, but 6–12% of capacity is lost per day.

The hydrogen electrode is generally a platinized or activated porous nickel mass which gives very good reversibility and low polarization. The nickel oxide electrodes are chemically or electrochemically impregnated with nickel hydroxide, as in the sintered plate nickel–cadmium cells. The battery generally consists of a stack of nickel oxide and hydrogen electrodes separated by zirconium oxide cloth or asbestos felt separators used to immobilize the 30% KOH aqueous electrolyte. A schematic diagram of the SAFT hydrogen–nickel system is shown in Fig. 9.23. The hydrogen electrode substrate has a central channel to facilitate gas diffusion.

Union Carbide showed the possibility of developing D-size hydrogen–nickel batteries with satisfactory performance. However, safety and cost considerations have restricted the applications of such units and the discovery of hydrogen storage alloys led to the development of the 'nickel–metal hydride' secondary cell which was described in Chapter 6.

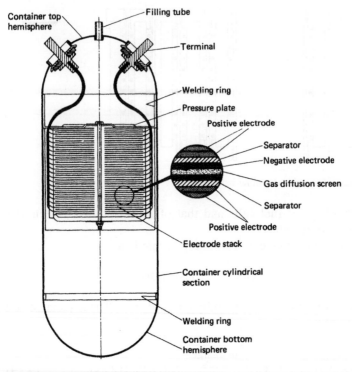

Fig. 9.23 Cross-section of a SAFT hydrogen–nickel battery. (By permission of Academic Press: S. Font and G. Goulard, *Power Sources* 5, ed. D. H. Collins, 1975, p. 334.)

Hydrogen–silver cells are attractive (apart from cost) since the energy density is higher than the hydrogen–nickel analogue. The cell is

$$Pt(s),H_2(g)|KOH(aq)|AgO(s)$$

and the cell reactions are basically

$$AgO(s) + H_2(g) \underset{\text{charge}}{\overset{\text{discharge}}{\rightleftharpoons}} Ag(s) + H_2O(l) \qquad (9.20)$$

The cell design is similar to that of the hydrogen–nickel system. There are problems with silver oxide solubility in the alkaline electrolyte, but charge retention is comparable with that of nickel–cadmium.

9.3   Redox cells

The use of two redox couples as active components has been proposed for cells designed for load levelling, bulk energy storage and more recently for electric vehicles. Such systems have inherently low energy densities and

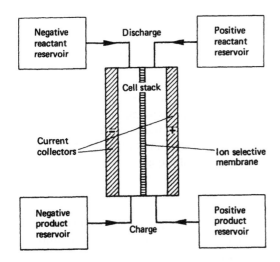

Fig. 9.24 Schematic cross-section of a redox flow cell

are not likely to be suitable for other applications. Two types of cell system are being considered.

Flow cells

This form of cell is shown schematically in Fig. 9.24. The anolyte and catholyte are different redox solutions which flow or are pumped past inert electrodes. The cell is constructed of two compartments separated by an anion-selective semi-permeable membrane. The spent solutions are retained in storage tanks and the whole process is reversed during charge. The general cell reaction is thus

$$A_{red}(aq) + B_{ox}(aq) \underset{\text{charge}}{\overset{\text{discharge}}{\rightleftharpoons}} A_{ox}(aq) + B_{red}(aq) \qquad (9.21)$$

The system has a number of attractive features, especially the flexibility of capacity and power output: capacity can be increased simply by enlarging the size (or number) of the storage tanks; power output can be raised by increasing the flow rate or by bringing into line parallel cell sections (cf. fuel cells). Further advantages include low operating temperatures and the absence of discharge depth or cycle life limitations.

A screening of possible redox couples has led to the choice of Fe^{3+}/Fe^{2+} as a very suitable catholyte, while Cr^{2+}/Cr^{3+} and Ti^{3+}/TiO^{2+} have been suggested for the anolyte. The latter shows problems connected with hydrolysis effects unless very low pH is maintained, and in addition has a low exchange current density. Cr^{2+}/Cr^{3+} is also kinetically slow, especially in the charging direction. In cells tested so far, voltages and current densities have been low and considerable improvements are needed in electrode performance and in the development of anion-selective membranes with lower resistivity. Theoretical models of 10 and 100 MWh systems have been developed by NASA/ERDA in the USA.

A redox battery using solutions with two different oxidation states of vanadium has recently been announced by the University of New South Wales in Sidney, Australia. When fully charged, each cell of the battery can generate a potential of about 1.5 V. A demonstration golf cart driven by the battery has been developed. Licences to industries in Thailand and Japan are underway for the production of large units for back-up power in solar houses or for peak demand in power stations.

Cells with insoluble redox couples

A system has been proposed which involves quinone/hydroquinone interconversions. A number of such couples are insoluble in acid solution and can therefore be used as active solid masses when made conductive by the addition of graphite. A system which has been studied experimentally is based on anthraquinone, Q, and tetrachloro-p-benzoquinone (chloranil), Q′, and their reduced or hydroforms. The overall cell reactions are then

$$H_2Q'(s) + Q(s) \underset{charge}{\overset{discharge}{\rightleftharpoons}} Q'(s) + H_2Q(s) \tag{9.22}$$

No solution flow or ion-selective membrane is required and the volume of electrolyte required is low. The total amount of acid is invariant since only proton transfers are involved. The OCV of the cell is 0.6 V and the theoretical energy density is 67 Wh/kg.

9.4   Thermal batteries

A thermal battery is a device which is able to supply energy only after it has been activated by the rapid application of heat. This condition can be achieved with systems based on electrolytes which change in a discontinuous manner from a low to a high conductivity state when heated. Such an electrolyte may typically be a solid with a phase transition to an ionically conductive phase (considered earlier in this chapter) or a salt mixture with a low melting point. For example, the LiCl–KCl eutectic mixture described in Chapter 8 is solid and chemically and electrochemically inert at room temperature, but becomes highly conductive on melting at 352°C. A thermal battery is thus a primary reserve system, which in its inert condition has a practically unlimited storage life, is usually constructed to be resistant to mechanical stress, humid environments etc., but which is capable of supplying a single high power discharge whenever required. The discharge pulse may be designed to last from a few tens of seconds up to an hour. Thermal activation is generally obtained using in-built pyrotechnic heat

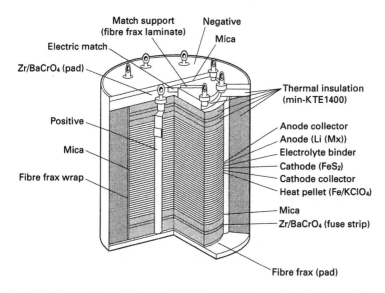

Match support
(fibre frax laminate)   Negative
                          Mica
Electric match
Zr/BaCrO₄ (pad)

Positive
Mica
Fibre frax wrap

Thermal insulation
(min-KTE1400)

Anode collector
Anode (Li (Mx))
Electrolyte binder
Cathode (FeS₂)
Cathode collector
Heat pellet (Fe/KClO₄)

Mica
Zr/BaCrO₄ (fuse strip)

Fibre frax (pad)

Fig. 9.25 Cross-section of 0.5 A, 28 V lithium–iron disulphide 60 minute thermal battery. The battery contains 15 active cells and has a total volume of 400 cm³. (By courtesy of Sandia National Laboratories.)

sources ignited by an electrical signal, and the electrolyte can be melted in a few tenths of a second (1–3 s for larger batteries). Thermal batteries are used mainly for military, aerospace and emergency applications.

A practical lithium–iron disulphide thermal battery developed by Sandia National Laboratories is shown in Fig. 9.25. This typical reserve battery uses the LiCl–KCl eutectic as electrolyte, LiSi alloy as anode material and FeS₂ as cathode. (The electrochemical characteristics of these components were described in Chapter 8.) The anode is prepared by directly pressing finely powdered lithium–silicon alloy into a cylindrical pellet. The electrolyte pellet is formed similarly using a mixture of LiCl–KCl eutectic and MgO binder (65%–35%) – the latter immobilizes the electrolyte when molten and acts as a separator. The cathode pellet is a mixture of 64% FeS₂, 16% LiCl–KCl eutectic and 20% electrolyte binder (here SiO₂). The heat pellet associated with each cell is a mixture of 88% iron powder and 12% potassium perchlorate. This pellet is slightly larger in diameter than the others, to improve contact with the fuse ignition strip (formed by a slurry of Zr/BaCrO₄ mixture). The anode and cathode current collectors are stainless steel discs, which also act as thermal buffers and prevent anode and cathode materials from reacting with the heat pellets.

Cells are assembled in the following sequence: anode collector, anode pellet, electrolyte pellet, cathode pellet, cathode collector and heat pellet. The battery is formed by pressing together 15 of these single cells. The assembly is housed in a closed cylindrical container provided with good thermal insulation and glass-to-metal sealed top feedthroughs for the external leads and for the electric firing signal which ignites the fuse strip.

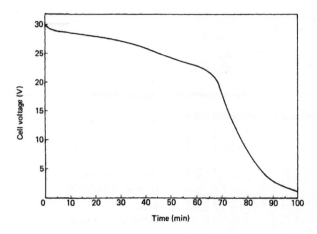

Fig. 9.26 Constant load discharge of lithium–iron disulphide thermal battery. Load resistance = 56 Ω. (By courtesy of Sandia National Laboratories.)

When activated, the battery supplies the typical discharge current shown in the discharge curve in Fig. 9.26. The beginning of the discharge in this battery, and more generally in all thermal batteries, is characterized by a peak voltage which falls off as the internal resistance of the cell rises, as a result of a build-up of insoluble products at the electrodes and a fall in temperature of the electrolyte. Usually the life of a thermal battery is measured as the time elapsed to reach a fixed fraction (e.g. 80%) of the peak voltage.

Other systems

Other common anode materials for thermal batteries are lithium alloys, such as Li/Al and Li/B, lithium metal in a porous nickel or iron matrix, magnesium and calcium. Alternative cathode constituents include $CaCrO_4$ and the oxides of copper, iron or vanadium. Other electrolytes used are binary KBr–LiBr mixtures, ternary LiF–LiCl–LiBr mixtures and, more generally, all lithium halide systems, which are used particularly to prevent electrolyte composition changes and 'freezing out' at high rates when lithium-based anodes are employed.

The calcium–calcium chromate thermal cell has been established for many years. In the LiCl–KCl eutectic, the reaction product of this cell is a mixed lithium–calcium–chromium oxide. However, this system cannot provide as high a specific capacity or energy density as the lithium-based systems described above. Furthermore, it suffers from parasitic chemical reactions which are exothermic and often uncontrolled.

9.5 Supercapacitors

Introduction

Electrolytic capacitors are common components in electric circuits. Special types designed to provide electrical pulses at high power levels (in the kW/kg range) for very short times (e.g. in 10–20 s) are being developed for special applications. One relevant example is the use of such capacitors in combination with a battery in an electric vehicle, where the capacitors can provide the power peaks needed for acceleration or hill climbing, while the main battery operates in a lower power, continuous regime. This combined action improves the reliability and longevity of the EV power system while providing improved performance.

These high power (and high capacitance) devices are commonly termed 'supercapacitors'. In its essential structure, a supercapacitor is very similar to an electrochemical cell: both are formed by two electrodes separated by an electrolyte. The difference is that the former operates on the basis of a Faradaic process and the latter does not. In other words, while in a battery the charge–discharge process is accompanied by the exchange of electrons between electrode and electrolyte, thus involving changes of oxidation states of the cell components, in a capacitor the process involves only the charging and discharging of the double layer at the electrode/electrolyte interfacial region. This difference is reflected in the values of energy and power density. A supercapacitor has an energy density which, although higher than that of conventional capacitors, is still much lower than that of advanced batteries, e.g. typically 10–15 Wh/kg in comparison with 80–100 Wh/kg. On the other hand, the absence of phase transformations, and thus of kinetic polarization limitations, allows supercapacitors to have much higher power density and longer cycle life than batteries, e.g. typically 500 W/kg and 10^5 cycles, in comparison with, say, 200 W/kg and 500 cycles.

Two types of electrochemical supercapacitors have been developed and tested. One operates as outlined above, namely by charging and discharging the interfacial double layer. In the second type, often called 'pseudo-capacitors' or 'redox capacitors', the charge–discharge mechanism involves the transfer of electronic charge between the phases, but without the occurrence of phase transformations. These processes include selective occupation of specific crystallographic sites, changes of oxidation states on the surface of metal oxides or the formation of electronic charges in highly delocalized states, such as on polymer backbones.

Double-layer supercapacitors

The amount of charge which can be stored in the electrode/electrolyte double layer is typically of the order of 15–40 $\mu F/cm^2$. A large capacitance may be obtained using electrode materials produced in highly dispersed form which have a very large surface area.

The capacitance is given by

$$C = \varepsilon_0 \varepsilon \, \frac{A}{d} \qquad (9.23)$$

where ε_0 is the absolute permittivity of free space, ε is the relative permittivity of the region between the electronic conductor and the inner Helmholtz plane (Fig. 2.2), A is the electrode surface area and d is the distance between the conductor and the inner Helmholtz plane. The most successful approach to fabrication of practical supercapacitors has been based on selecting carbon powders of various types and specifications. For instance, useful carbon electrodes can be obtained by pyrolysis of a phenolic resin at a temperature ranging from 800 to 1100°C. Following this process, the active surface area (A in eq. 9.23) may increase up to 250 times. Other types of electrodes have been prepared by sintering carbon–metal mixed fibres. This process gives electrodes which have a high surface combined with low resistance, and thus high capacitance combined with high power. Carbon electrodes having surface areas in the range of 1000–2000 m²/g are currently available for double-layer supercapacitor use.

Fig. 9.27 shows the variation of capacitance as a function of the surface area from which it may be concluded that in principle it is possible to obtain double-layer supercapacitors having specific capacitance of the order of 100 F/g so that values of practical capacitance density as high as 30 F/g can be predicted.

A typical configuration of a double-layer supercapacitor involves two metallic collectors which hold in place the carbon powder electrodes, which in turn are separated by an electrolyte, in most of the cases formed by liquid solutions (Fig. 9.28.) A layer of porous, non-conductive material acts as a separator.

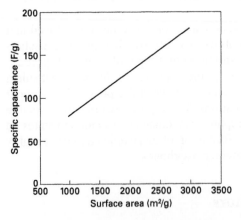

Fig. 9.27 Specific capacitance of a double-layer supercapacitor as a function of surface area. (By permission of *Chim & Ind.*: E. Spila *et al.*, 1995, **77**, 333.)

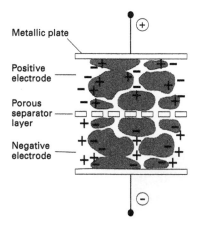

Metallic plate

Positive electrode

Porous separator layer

Negative electrode

Fig. 9.28 Schematic diagram of a double-layer supercapacitor. (By permission of *Chim & Ind.*, as Fig 9.27.)

The highest usable voltage of a supercapacitor is set by the electrolyte decomposition voltage. The electrolyte can be either aqueous (e.g. strong acidic or basic solutions) or organic (e.g. salt solutions in aprotic solvents). The former give higher capacities and higher power density (related to the lower resistance of the aqueous electrolytes, i.e. typically 10^{-1}–10^{-2} S/cm in comparison with 10^{-3}–10^{-4} S/cm for non-aqueous systems); the latter gives higher voltages (related to the higher electrolyte decomposition potential, i.e. about 3.5 V for the organic solutions in comparison with 1.2 V for the aqueous solutions) and thus, higher energy density.

Fig. 9.29 shows a plot of power density versus energy density for supercapacitors in comparison with some conventional and advanced batteries where it may be seen that supercapacitors typically operate in the very high power density range (i.e. 400–1000 W/kg) but with energy densities of only a few Wh/kg. This is confirmed by Table 9.2 which lists the characteristics of some prototypes presently under development.

Solid ionic conductors can also be used in the fabrication of solid state double-layer supercapacitors. An example is the device developed in the late 1960s by Gould Ionics which adopted a cell system using a silver–carbon electrode couple separated by the highly ionically conducting solid electrolyte $RbAg_4I_5$ (see Section 9.1):

$$Ag(s)|RbAg_4I_5(s)|C(s)$$

This solid state supercapacitor was fabricated by compressing electrode and electrolyte powders into a single pellet. The maximum operating voltage was set by the decomposition of the $RbAg_4I_5$ electrolyte which is 0.66 V. Typically, the device was rated for operation over the 0–0.5 V range and was designed as long time-constant timing element, as well as back-up for volatile memory. However, full commercialization was not pursued, probably because of cost and voltage limitations.

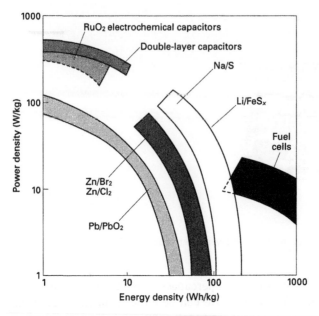

Fig. 9.29 Power and energy densities of double-layer and electrochemical supercapacitors in comparison with typical rechargeable batteries and fuel cells

Table 9.2 Characteristics of some double-layer capacitor prototypes

Company	Electrode material	Electrolyte	Energy density (Wh/kg)	Power density (W/kg)
FY	Carbon	H_2SO_4	0.33	—
FE	Carbon	H_2SO_4	0.01	—
Panasonic	Carbon	Organic	2.2	400
Evans	Carbon	H_2SO_4	0.2	—
Maxwell-Auburn	Composite C/metal	KOH	1.2	800
Maxwell-Auburn	Composite C/metal	Organic	7	2000
Livermore National Laboratory	C (aerogel)	KOH	1	—
Sandia National Laboratory	C (synthetic)	Aqueous	1.4	1000

In the late 1980s, the system was reconsidered by Quadri Electronics who produced an improved supercapacitor under the trade name HYPER-CAP™. Very high rate and peak power capabilities – current pulses in excess of 10 A with rise times of the order of milliseconds, and 3 kW/kg, respectively – have been reported for these solid state devices.

Redox electrochemical capacitors

In the second type of supercapacitor, sometimes termed pseudocapacitors, redox capacitors or electrochemical capacitors, the non-Faradaic double-layer charging process is accompanied by charge transfer. This Faradaic process must be characterized by extremely fast kinetics in order to allow device operation with high current density discharge pulses.

There are two classes of redox electrochemical capacitors, depending on the nature of the electrodes. One uses symmetrical pairs of metal oxide electrodes, generally based on RuO_x, IrO_x, TaO_x, MnO_x or their mixtures. The most common is ruthenium oxide, RuO_x, where x may vary from 1.9 to over 2.0. If properly prepared, RuO_x may have an effective surface area of the order of 100 m^2/g. The Faradaic part of the charging process of RuO_x-based electrochemical capacitors involves the partial oxidation-reduction of the electrodes involving the different oxidation states of ruthenium. These are fast processes and this, combined with the high surface area of the electrodes, confers high rate and high capacitance to the RuO_x electrochemical capacitors. The high price of the electrode material is, however, a significant drawback. The second class of electrochemical capacitor presently under consideration uses electronically conducting polymers such as poly(pyrrole), poly(thiophene) or poly(aniline), as the electrode-pair components. These polymers may be readily prepared, either chemically or electrochemically, in forms having high surface area morphologies. The Faradaic component of the charging process is the doping–undoping redox reactions of the polymers (see Chapter 7). Due to the high mobility of the electronic charges in the doped state of the polymers, these capacitors may combine high capacity with high rate capability and low cost.

There are three types of polymer-based electrochemical capacitors. The first uses a symmetric combination of the same p-dopable polymer (e.g. poly(pyrrole)) on each side. The charging process leads the electrodes to acquire a high doped state at the positive and a low doped state at the negative, with a total voltage difference of about 1 V. During discharge, the doping (i.e. oxidation) state of the electrodes varies in a complementary way to reach finally an intermediate value. A typical discharge curve is shown in Fig. 9.30. The specific capacity achievable with this type of capacitor is about 80 F/g. The second type uses an asymmetric system based on two different p-dopable polymers (e.g. poly(pyrrole) and poly (thiophene)). Since the two polymers are selected by virtue of their different doping potential, the system allows the voltage value of the fully charged device to be increased to about 1.5 V and also allows the amount of charge which can be released during discharge to be increased. Fig. 9.30 shows a typical discharge curve. The specific capacitance of these polymer capacitors is of the order of 90 F/g. The third type employs a symmetric system based on the same conducting polymer but which can be both p-doped (oxidized) and n-doped (reduced). Examples are poly-3(4 fluoro-phenyl)thiophene or poly(dithieno[3,4-B: 3′,4′D]thiophene. During charge,

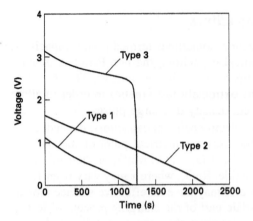

Fig. 9.30 Constant current (200 mA) discharge curves for the three types of polymer electrochemical capacitor

one polymer electrode becomes fully p-doped and the other fully n-doped. Thus, the maximum achievable voltage is further increased to about 3 V and the charge is increased to the maximum available doping level. A typical discharge curve, shown in Fig. 9.30, reveals that most of this charge is released at high voltage, which is a further advantage of this type of capacitor.

Polymer electrochemical capacitors have not yet reached large-scale commercial production, but significant exploitation is expected in the near future.

Appendix 1: Operational modes and charging techniques for secondary batteries

The optimum method for recharging a battery depends both on the type and characteristics of the cell system under consideration, and on its service regime. The prime requirement is for a recharging system which will provide a rapid and efficient charge, using equipment which is as simple and inexpensive as possible and which will avoid damaging or reducing the cycle life of the battery. At its simplest, constant voltage charging may use a statically controlled rectifier with some form of current limiter to prevent excessive currents and Joule heating early in the charge. At its most sophisticated, the Smart Battery concept being developed by IBM incorporates a dedicated computer chip which undertakes power management, monitors charge levels and optimizes the charging of each individual cell in a battery. There are three basic patterns of service for secondary cells – cyclic operation, standby power application and continuous float or buffer mode operation.

Cyclic operation

In applications such as load levelling, motive power or in consumer electronics, the battery is required to undergo a sequence of deep discharges followed by a recharge to maximum capacity. A number of methods may be used.

Constant current

With this method, control equipment is simple and inexpensive and it is straightforward to calculate the charge delivered. However, if high currents are used, problems are encountered with batteries which have porous electrode structures because of non-uniform current distribution, and severe gassing may occur towards the end of the charge. At low current, the total charging time may be unacceptably long. This method can, however, be used with nickel–cadmium cells, which can sustain prolonged overcharging at C/10 without appreciable deterioration. In the simplest method, the current is controlled by a resistor in series with the battery and a DC power supply. 'Fast charging' methods are now more common; these require circuits to control the voltage and cut off the current at the end of charge. These techniques are considered in the section on automatically regulated systems below.

Constant voltage

With constant voltage charging, the current is high at the beginning of the charge and falls off as the charging proceeds, so that the current flow towards the end of the process is likely to be very low. If the charging voltage is too high, Joule heating early in the charge is likely to be excessive, leading to rapid deterioration of the cell.

Under certain circumstances, constant voltage charging of sealed secondary batteries can result in 'thermal runaway'. In this phenomenon, at the end of the normal charge, overcharge reactions take over and the energy of the charging current is largely converted to heat. The consequent rise in temperature of the cell lowers the overall polarization and this increases the charging current, which in turn may raise the temperature further, and so on until an explosion may result.

Automatically regulated systems

There is now a wide range of regulated charging systems available which use techniques such as step-wise reducing currents, constant voltage with current limitation, voltage taper charging, combinations of constant current and constant voltage regimes, taper pulse charging etc. These procedures have been developed for individual cell systems to minimize charging time, while avoiding excessive side reactions towards the end of charge. One of the simplest of these methods is taper charging. Basic taper chargers have a constant high supply voltage and use a simple series resistor to modify the fall-off in current as the battery voltage rises. An example of a more sophisticated electronically controlled taper charger is the 'Spegel' system (Chloride Industrial Batteries) for lead–acid cells, which delivers a current that is inversely proportional to the cell voltage. The OCV is monitored every 15 minutes during a current interruption. If the voltage is below 2.3 V, the charge continues normally (Fig. A1.1), but if it is higher, further application of charge is delayed. When the system recognizes a fully charged condition, short current pulses are supplied to retain the battery in a fully charged state.

Sealed nickel–cadmium and nickel–metal hydride sealed cells for consumer equipment now generally employ charge control circuits which permit fast charging. A number of methods are used to detect the onset

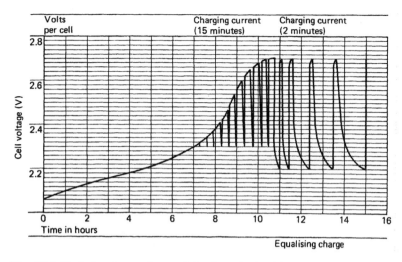

Fig. A1.1 Typical recharge characteristics for a lead–acid cell under Spegel control. (By permission of Chloride Industrial Batteries.)

of overcharge. For example, an increase in cell temperature due to oxygen generation at the positive generally leads to a detectable reduction in the cell voltage, and the detection of this voltage drop triggers the termination of the charge. This is known as the $-\Delta V$ method. Alternatively, by monitoring the rate of temperature increase, the $\Delta T/\Delta t$ method stops the charging current when the rate of temperature rise exceeds a certain value. Even more sophisticated chargers are available which can 'condition' a cell by fully discharging it to eliminate any memory effect (see p. 175), charge it at a high rate (e.g. 1–2 C) controlled by the $-\Delta V$ method, then add a 'top-up' charge using a low rate timed current pulse, and finally apply a maintenance charge at constant potential.

Rechargeable alkaline manganese batteries are charged using a repetitive current pulse method in which the OCV is measured during each period when the pulse is 'off' (Fig. A1.2). Once the OCV has reached 1.65 V, pulses are switched off until the OCV decreases. As the cell charges, a greater and greater proportion of the pulses are blocked, so that the average current tapers to zero (Fig. A1.2b).

Lithium and lithium ion cells are generally much less tolerant of overcharge than aqueous systems. Charging methods generally use some form of constant or taper current charging to a fixed maximum voltage. Particular care is needed when using such cells in series, where 'Smart Battery' computer monitoring and control is likely to become the norm.

Standby power

Here the consumer circuit is either connected to the mains supply and a battery-powered inverter and transformer system, or the latter supplies

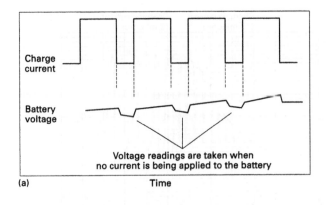

(a)   Time

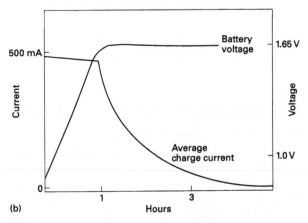

(b)   Hours

Fig. A1.2   (a) Pulse charge and OCV sensing of a RENEWAL rechargeable alkaline manganese cell. (b) Typical current and voltage profiles during charging of an AA sized RENEWAL rechargeable alkaline manganese cell using the pulse taper charge method. (By permission of Rayovac.)

power by means of a relay which closes in the event of mains failure (*see* Fig. 1.5). In either case, the battery is only required to deliver current very occasionally during its entire working life. The normal charging requirement is therefore solely to compensate for self-discharge and thus to maintain the system at a state of full charge. This can be achieved by the technique known as trickle charging: the correct current input is one which exactly compensates for self-discharge and does not lead to gassing. Various empirical formulae are used for calculating the appropriate current level: for example, for lead–acid cells with Pb–Se grids a current of 1 mA/ Ah of nominal capacity at C/10 is recommended for new cells, rising to 3 mA/Ah towards the end of cell life. It is necessary to incorporate in standby power installations circuits capable of high current (boost) charging following a deep emergency discharge. Standby batteries can also be float charged – i.e. battery, load and constant voltage charging circuit are connected in parallel, and the latter can deliver sufficient output to meet the normal load demand plus any charging requirements. For lead–acid cells the voltage used is commonly 2.25 ± 0.05 V at ambient temperature. For very low maintenance systems (typically nickel–cadmium cells) where a long interval between electrolyte topping up is required, it is common to operate cells well below their fully charged state, e.g. nickel–cadmium cells may be float charged at 1.42 V.

Continuous float or buffer mode operation

Again the consumer circuit is connected in parallel with the battery and a generator, which in this case may be the alternator or dynamo of a car engine, a wind-driven system, solar power cells, etc. If the load demand is small, or the generator is producing a high power level, the supply bus voltage rises and the battery can accept charge. For peak loads, or times when the generator power falls, the battery augments the supply. The battery must be protected from gross overcharging caused by high voltage excursions in the generator, so that voltage control devices are necessary. Further, suitable diodes must be inserted to prevent discharge of the battery through the generator. Such a system can only work successfully if the generator can supply sufficient energy to recharge the battery fully during the normal working sequence, i.e. the long-term time integral of the power demand of the consumer circuit, together with any battery losses, must be less than that of the generator output over the same period. The selection of a battery with the correct capacity and appropriate rate characteristics is of great importance in this type of application.

It must be emphasized that the most appropriate charging regime is very dependent on the cell system under consideration. Some are tolerant to a considerable amount of overcharging (e.g. nickel–cadmium batteries), while for others, such as zinc–silver oxide and most lithium secondary cells, overcharging can result in permanent damage to the cell. Sealed battery systems require special care: float charging should not be used and trickle charge rates should be strictly limited to the manufacturer's recommended values, since otherwise excessive cell temperatures or thermal runaway can result.

Appendix 2: Nomenclature and standardization systems for small cells and batteries

There is still considerable confusion between competing nomenclatures. Because of the increasing importance of the world trade in batteries, however, international standards for specification, testing and nomenclature are now becoming more significant. Unfortunately, many manufacturers adopt their own systems of identifying their cells, and cross-reference charts are often required. While the traditional letter designations for cylindrical cells summarized in Fig. 1.4 are still widespread, a more informative nomenclature is being successfully promoted by the International Electrotechnical Commission (IEC), the organization responsible for worldwide standardization in the electronics and electricity industries. IEC standards currently extend to the most important primary cells (publications IEC 86-1, 86-2 together with two sets of amendments), nickel–cadmium and lead–acid batteries. The objectives of their standards are to ensure the interchangeability of cells from different manufacturers, from both an electrical and a size point of view. So far as primary batteries are concerned, the IEC nomenclature specifies:

- the number of cells in series and/or parallel;
- the electrochemical system, and by inference the OCV;
- the shape;
- the nominal dimensions.

The system is designated by a letter, e.g. 'B' specifies $Li/(CF_x)_n$, 'C' specifies Li/MnO_2, 'L' the Zn/MnO_2 alkali cell, etc. (A 'blank' rather than a letter is used for Leclanché or zinc–carbon cells.) The shape is also specified by a letter: 'R' for cylindrical, 'S' for prismatic and 'F' for flat. The following examples illustrate how the system works.

Example 1

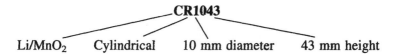

$$CR1043$$

Li/MnO$_2$ Cylindrical   10 mm diameter   43 mm height

To simplify the designation of dimensions, some well-established cells are given codes – e.g. the cylindrical AA-cell is known simply as R6.

Example 2

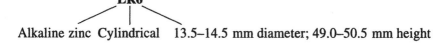

LR6

Alkaline zinc Cylindrical   13.5–14.5 mm diameter; 49.0–50.5 mm height

Finally, the number of cells in series is given by a prefix, and the number of cells (or series groups) in parallel by a hyphenated suffix.

Example 3

4LR6-2
Two parallel groups of four LR6 cells in series

It should be noted that specifications, regulatory and safety standards are promulgated by a wide variety of organizations, including national standards committees, military forces, transportation authorities and insurance underwriters.

Appendix 3: Waste disposal and the recycling of batteries

At the present time, a large number of spent batteries are disposed of directly into the urban waste stream without proper controls. In addition to the most common systems such as zinc–carbon, alkaline manganese and nickel–cadmium, these now include, at an increasing rate, nickel–metal hydride and lithium cells. Such disposal is of serious concern because of the possible effects of battery components on the environment. Consequently, most countries are now evolving policies for collection and recycling. The majority of lead–acid batteries are recycled, but the number of recycling plants in operation worldwide for other battery systems is still very small due to the unfavourable economic balance of such operations (see Table A3.1). Some of the procedures for the disposal and recycling of battery materials are now briefly described.

Primary aqueous cells ('dry batteries')

As is well known (and discussed in Chapter 3), traditional dry batteries use zinc as the negative electrode. While this metal offers many advantages as an electrode material, it is attacked by most common electrolytes and thus

Table A3.1 Status of present technology for recycling spent batteries

Battery type	Materials recovered	Fraction of cells recycled	Economic advantages
Zinc-based primary cells	Zn, Hg Fe–Mn	Minor	Negative
Lead–acid batteries	Pb, plastics	>90%	Good
Nickel–cadmium batteries	Cd	50% (industrial) Minor (consumer)	Acceptable

corrosion prevention is an important aspect of dry cell technology. Until recently, mercury and sometimes small quantitites of cadmium were added to the anode to increase the hydrogen overvoltage and thus to protect the zinc from corrosion. The environmental hazard of these heavy metals is now well known and most major battery producing companies have attempted to reduce or eliminate the heavy metal content from dry batteries sold on the consumer market. Cells without any heavy metal are now labelled as 'mercury-free' or 'green batteries'.

Although the reduction or elimination of mercury is important, it is not the case that all spent dry batteries are now suitable for direct insertion into the solid urban waste stream. Hence, following a European Community directive (91/157) in 1991, many western European countries have listed such batteries in the category of hazardous waste. In Germany, Scandinavia and Italy, considerable efforts are being made to inform consumers of the environmental danger of spent batteries and to introduce better methods of collection. Although this campaign has now been in force for some time, the results so far have not been very encouraging since collection amounts to only about 20–30% of total battery sales. In some other European countries, notably the UK, implementation of the directive has been at best half-hearted.

Collection is, however, only one aspect of the procedure: disposal and recycling remain the most critical, and often unsolved, parts of the battery waste cycle. Recycling plants are still very scarce. One of the first, developed by Sumitomo Heavy Industries Ltd in Japan, is based on a pyrometallurgic process. Used dry batteries (UDBs) are heated in a furnace, initially at 300–700°C where the majority of any mercury present is reduced to the metallic state, organic materials are pyrolysed into combustible volatile matter and a solid coke, and any halogen-based components are partially decomposed. The residue is then fed into a high temperature furnace (1500°C) where residual zinc, iron and manganese oxides are melted and reduced to zinc vapour and a ferromanganese alloy. (Note that the carbon electrodes of the batteries act as reducing agent.) The zinc vapour is introduced into a condenser from which metal is recovered in the form of ingots. The ferromanganese alloy and mercury are also recovered. A series of purification and control units ensure that any gas and liquid effluents are mercury-free prior to discharge.

There are now two plants operating in Europe, both located in Switzerland. The plant developed by Batrec AG in Wimmis has been in operation since 1994 and has a nominal capacity of 2000 tonnes of UDBs per annum. This plant is based on the Sumitomo process; its flow sheet is shown schematically in Fig. A3.1.

The second Swiss plant was developed by Recycled SA in Dietikon, and also commenced operation in 1994. It uses a thermal treatment of UDBs aimed at achieving full de-mercurization. Although effective, these recycling plants are based on highly expensive processes and thus are not profitable in strict monetary terms.

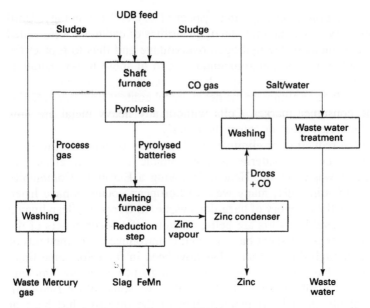

Fig. A3.1 Flow sheet for the Sumitomo dry battery recycling process. (By permission of *J. Power Sources*: R. Burry and A. Weber, 1995, **57**, 31.)

Lead–acid batteries

In the case of lead–acid batteries, recycling of exhausted units is under-taken worldwide and the process is both efficient and cost-effective. It has been calculated that almost 90% of spent lead–acid batteries are sent back to recycling plants. This high return is explained by the large scale of lead–acid cell production, which makes recycling mandatory in order to control the price of lead on the world market.

Most of these units adopt physical or chemical processes to separate the components and then thermal treatment for smelting and refining. The components are polypropylene (from the cases), lead and lead compounds (from the grids, terminals and paste slurry), acid (from the electrolyte) and other residues (separators, fibres, etc.). Smelting is typically conducted in furnaces designed to produce crude lead. Further refining is used to synthe-size a range of alloys to meet specific mechanical, electrical and chemical characteristics. Electrochemical processes are occasionally used.

Cadmium-based batteries

Despite the fact that cadmium is very toxic and that the number of batteries based on this metal (i.e. nickel–cadmium cells) is very significant, there are very few recycling plants in operation. One very effective recycling pro-cess has been designed and developed by SAFT NIFE AB in Sweden to treat both industrial and consumer nickel–cadmium batteries. The flow

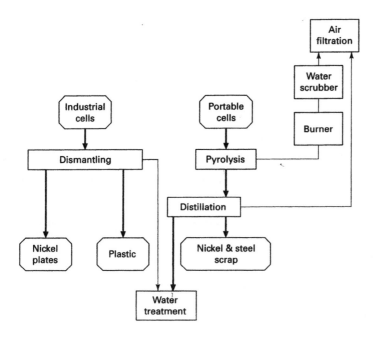

Fig. A3.2 Flow sheet for the SAFT NIFE AB nickel–cadmium cell recycling process.

sheet is shown in Fig. A3.2. Industrial cells are dismantled and their negative plates (which contain the cadmium) are washed, dried and heated with coke at 900°C in an electric furnace for 24 hours. Cadmium is then distilled off and the liquid metal cooled in water to form small metallic pellets. Sealed nickel–cadmium batteries are subjected to pyrolysis prior to distillation in order to decompose any polymer compounds. The pyrolysis gas is completely combusted, washed and filtered before being emitted into the atmosphere.

A similar process developed in the 1980s by the Inmetc Co. in the United States is capable of handling about 3000 tonnes of industrial and consumer nickel–cadmium batteries per annum. Here again, the cadmium is distilled and converted into ingots which are supplied back to the battery industry. Finally, another thermal process for treating nickel–cadmium battery waste is in operation in France by the Societé Nouvelle d'Affinage des Metaux (SNAM).

Lithium batteries

The entry of lithium batteries into the consumer market is still quite recent and not much attention has so far been paid to disposal and the possibility of recycling procedures. Lithium metal, although not toxic, is a safety hazard since it is very reactive, especially in contact with water or in high humidity. This makes lithium batteries which have not been fully discharged of potential high risk, especially if fractured or dismantled without care. Further, concern has been expressed about the organic solvents which are used to constitute the electrolyte and the cathodes such as $LiCoO_2$ which are based on toxic metals (see Chapter 7).

The main problem to be faced in the organization of a lithium battery waste process is to eliminate any residual lithium metal by forced discharge of the cell. Other appropriate steps would include:

- draining any toxic liquid electrolyte or liquid cathode and their neutralization in *ad hoc* chemical treatment units;

- remote sectioning of the lithium-containing portion of the batteries and final deactivation of any residual lithium.

Some attempts to develop waste processing units using these concepts are in progress in the United States where proper disposal of large military high power and high capacity cells is now necessary. There are not yet comparable efforts in Europe, and no lithium battery disposal plants for consumer cells have yet been planned.

Appendix 4: Electrical quantities, physical constants and conversion factors

A4.1 Electrical quantities

Potential difference; emf V (volt)
Electrical field (gradient) V/m (volt per metre)
Resistance Ω (ohm)
(Specific) resistance Ωm (ohm metre)
Conductance S (siemens); 1 S = 1 Ω^{-1}
(Specific) conductance S/m (siemens per metre)
Electric charge C (coulomb)
Current A (ampere); 1 A = 1 C/s
Current density A/m^2 (ampere per square metre)
Capacity Ah (ampere hour); 1 Ah = 3600 C
    [NB. The capacity of a battery (Section 2.5) measured in Ah should not be confused with the *capacitance* of a capacitor (condenser) which is measured in F (farads)]
Capacity density or specific capacity Ah/kg (ampere hour per kilogram)
Volumetric specific capacity Ah/dm^3 (ampere hour per cubic decimetre)
Power W (watt); 1 W = 1 J/s
Power density or specific power W/kg (watt per kilogram)
Volumetric power density W/dm^3 (watt per cubic decimetre)
Energy; work J (joule); 1 Wh = 3.6 kJ
                 1 J = 1 C.V
Energy density or specific energy Wh/kg (watt hours per kilogram);
                        1 Wh/kg = 3.6 kJ/kg
Volumetric energy density Wh/dm^3 (watt hours per cubic decimetre);
                    1 Wh/dm^3 = 3.6 kJ/dm^3

A4.2   Physical constants

Avogadro number (L) 6.022×10^{23} molecules per mole
Boltzmann's constant (k) 1.381×10^{-23} J/K per molecule
Faraday's constant (F) 9.649×10^4 C per equivalent
Planck's constant (h) 6.622×10^{-34} J s
Charge on elctron (e) 1.601×10^{-19} C
Gas constant (R)   8.314 J/K per mole

A4.3   Conversion factors

Length	1 inch	0.0254 m
	1 foot	0.3048 m
Area	1 sq. inch	6.45×10^{-4} m^2
	1 sq. foot	0.0929 m^2
Volume	1 cu. inch	1.639×10^{-5} m^3
	1 cu. foot	0.0283 m^3
	1 gallon (UK)	0.004 55 m^3
	1 gallon (US)	0.003 79 m^3
Mass	1 tonne	1000 kg
	1 lb	0.4536 kg
	1 ton (long)	1016 kg
Force	1 dyne	1.000×10^{-5} N
Pressure	1 atmosphere ⎱	1.013×10^5 Pa
	760 mmHg ⎰	
	1 torr	1.333×10^2 Pa
	1 bar	1.000×10^5 Pa
Power	1 hp (horsepower)	746 W
Energy	1 eV	1.602×10^{-19} J
	1 erg	1.000×10^{-7} J
	1 cal	4.184 J
	1 Wh	3600 J
	1 BTU (British thermal unit)	1055 J (0.2931 Wh)
	1 mtoe (million ton oil equivalent)	4.187×10^{15} J (1.163×10^{13} Wh)
Charge	1 esu (statcoulomb)	3.333×10^{-10} C

A4.4   SI prefixes for submultiples and multiples

a	(atto)	10^{-18}	da	(deka)	10^1
f	(femto)	10^{-15}	h	(hecto)	10^2
p	(pico)	10^{-12}	k	(kilo)	10^3

n	(nano)	10^{-9}	M	(mega)	10^{6}
μ	(micro)	10^{-6}	G	(giga)	10^{9}
m	(milli)	10^{-3}	T	(tera)	10^{12}
c	(centi)	10^{-2}	P	(peta)	10^{15}
d	(deci)	10^{-1}	E	(exa)	10^{18}

Bibliography

General

G. Bianchi and T. Mussini, *Electrochimica*, Masson, Milan, 1993.

J.O'M. Bockris and A.K.N. Reddy, *Modern Electrochemistry*, Volumes 1 and 2, Macdonald Technical and Scientific, London, 1970.

J. Koryta, J. Dvorak and L. Kavan, *Principles of Electrochemistry*, 2nd edition, Wiley, Chichester, 1993.

D.B. Hibbert, *Introduction to Electrochemistry*, MacMillan, Basingstoke, 1993.

G. Kortum, *Treatise on Electrochemistry*, Elsevier, Amsterdam, 1965.

P.H. Rieger, *Electrochemistry*, Prentice-Hall, New Jersey, 1987.

W.M. Latimer, *Oxidation Potentials*, Prentice-Hall, N.J., 1952.

K.S. Pitzer and L. Brewer, *Thermodynamics* (G.N. Lewis and M. Randall), McGraw-Hill, New York, 1961.

M. Pourbaix, *Atlas d'Equilibres Electro-Chimiques*, Gauthier-Villars, Paris, 1963.

W.J. Albery, *Electrode Kinetics*, Clarendon Press, Oxford, 1975.

H.H. Bauer, *Electrodics*, Thieme, Stuttgart, 1972.

P. Delahay, *Double Layer and Electrode Kinetics*, Wiley, New York, 1965.

H.R. Thirsk and J.A. Harrison, *Guide to the Study of Electrode Kinetics*, Academic Press, London, 1972.

K.J. Vetter, *Electrochemical Kinetics*, Academic Press, London, 1967.

R.N. Adams, *Electrochemistry at Solid Electrodes*, Dekker, New York, 1969.

P.A. Christensen and A. Hamnett, *Techniques and Mechanisms in Electrochemistry*, Blackie, Glasgow, 1994.

E. Gileadi, *Electrode Kinetics*, VCH, New York, 1993.

J.P. Hoare, *Electrochemistry of Oxygen*, Wiley, New York, 1968.

A.T. Kuhn, *The Electrochemistry of Lead*, Academic Press, London, 1979.

D. Pletcher, *A First Course in Electrode Processes*, The Electrochemical Consultancy, Romsey, 1991.

D.H. Everett and F.S. Stone, *Structure and Properties of Porous Materials*, Butterworth Scientific, London, 1958.

S.J. Gregg and K.S.W. Sing, *Adsorption, Surface Area and Porosity*, Academic Press, London, 1967.

T. Erdey-Gruz, *Transport Phenomena in Aqueous Solutions*, Adam Hilger, London, 1974.

V.C. Levich, *Physicochemical Hydrodynamics*, Prentice Hall, N.J., 1962.

H. Harned and B. Owen, *The Physical Chemistry of Electrolytic Solutions*, Reingold, New York, 1958.

R.A. Robinson and R.H. Stokes, *Electrolyte Solutions*, Butterworth Scientific, London, 1959.

G. Charlot and B. Tremillon, *Chemical Reactions in Solvents and Melts*, Pergamon, Oxford, 1969.

A.K. Covington and T. Dickinson, *Physical Chemistry of Organic Solvent Systems*, Plenum Press, New York, 1973.

G.J. Janz and R.P.T. Tomkins, *Nonaqueous Electrolytes Handbook*, Volumes 1 and 2, Academic Press, London, 1972–3.

T.C. Waddington, *Nonaqueous Solvent Systems*, Academic Press, London, 1965.

Y.K. Delimarskii and B.F. Markov, *Electrochemistry of Fused Salts*, Sigma Press, Wolverhampton, 1961.

G.J. Janz, *Molten Salts Handbook*, Academic Press, London, 1967.

G. Mamantov, *Molten Salts: Characterisation and Analysis*, Dekker, New York, 1969.

G. Morand and J. Hladik, *Electrochimie des Sels Fondus*, Masson, Paris, 1969.

M. Balkanski, T.T. Takahashi and H.I. Tuller, *Solid State Ionics*, North Holland, Amsterdam, 1992.

P.G. Bruce, *Solid State Electrochemistry*, Cambridge University Press, Cambridge, 1995.

B.V.R. Chowdari, M. Yahaya, I.A. Talib and M.M. Salleh, *Solid State Ionics Materials*, World Scientific, Singapore, 1994.

C. Deports, M. Duclot, P. Fabry, J. Fouletier, A. Hammun, M. Kleitz, E. Siebert and J-L. Souquet, *Electrochimie des Solides*, Presses Universitaires de Grenoble, Grenoble, 1994.

S. Geller, *Solid Electrolytes*, Springer, Berlin, 1977.

P. Hagenmuller and W. van Gool, *Solid Electrolytes*, Academic Press, London, 1978.

A.L. Laskar and S. Chandra, *Superionic Solids and Solid Electrolytes*, Academic Press, Boston, 1989.

A.P. Legrand and S. Flandrois, *Chemical Physics of Intercalation*, Plenum Press, New York, 1987.

G.-A. Nazri, J.-M. Tarascon, and M. Armand, *Solid State Ionics III*, MRS Symposium Proceedings, Vol. 293, Pittsburg, Pennsylvania, 1993.

G.-A. Nazri, J.-M. Tarascon, and M. Schreiber, *Solid State Ionics IV*, MRS Symposium Proceedings, Vol. 369, Pittsburg, Pennsylvania, 1995.

B. Scrosati, A. Magistris, C.M. Mari and G. Mariotto, *Fast Ion Transport in Solids*, Kluwer, Dordrecht, 1992.

T. Takahashi, *High Conductivity Solid Ionic Conductors – Recent Trends and Applications*, World Scientific, Singapore, 1989.

T. Takahashi and A. Kozawa, *Application of Solid Electrolytes*, J.E.C. Press, Cleveland, 1980.

W. van Gool, *Fast Ion Transport in Solids, Solid State Batteries and Devices*, North Holland, Amsterdam, 1973.

P. Vashishta, J.N. Mundy and G.K. Shenoy, *Fast Ion Transport in Solids: Electrodes and Electrolytes*, Elsevier North Holland, New York, 1979.

F.M. Gray, *Solid Polymer Electrolytes: Fundamentals and Technological Applications*, VCH, New York, 1991.

R.G. Linford, *Electrochemical Science and Technology of Polymers – 2*, Elsevier, London, 1990.

J.R. MacCallum and C.A. Vincent, *Polymer Electrolyte Reviews I*, Elsevier, London, 1987.

J.R. MacCallum and C.A. Vincent, *Polymer Electrolyte Reviews II*, Elsevier, London, 1989.

B. Scrosati, *Applications of Electroactive Polymers*, Chapman and Hall, London, 1993.

Batteries

J.R. Akridge and M. Balkanski, *Solid State Microbatteries*, NATP ASI Series, Plenum Press, New York, 1990.

V.S. Bagotzky and A.M. Skundin, *Chemical Power Sources*, Academic Press, London, 1980.

M. Barak, *Electrochemical Power Sources*, Peter Peregrinus, Stevenage, 1980.

P. Bauer, *Batteries for Space Power Systems*, NASA, Washington DC, 1968.

R.V. Bobker, *Zinc-in-Alkali Batteries*, The Society for Electrochemistry/Joseph Lucas Ltd, University of Southampton, 1973.

J. O'M. Bockris, B.E. Conway, E. Yeager and R.E. White, *Electrochemical Energy Conversion and Storage*, Comprehensive Treatise of Electrochemistry, Vol. 3, Plenum Press, New York, 1981.

H. Bode, *Lead–Acid Batteries*, Wiley, New York, 1977.

P. Bro and S.C. Levy, *Quality and Reliability Methods for Primary Batteries*, Wiley, New York, 1990.

R.M. Dell, 'Advanced Secondary Batteries: A Review', UK Atomic Energy Authority, Harwell, 1979.

S. Falk and A.J. Salkind, *Alkaline Storage Batteries*, Wiley, New York, 1969.

A. Fleischer and J.J. Lander, *Zinc–Silver Oxide Batteries*, Wiley, New York, 1971.

J.-P. Gabano, *Lithium Batteries*, Academic Press, London, 1983.

A.G. Garrett, *Batteries of Today*, Research Press, Dayton, Ohio, 1957.

D.P. Gregory, *Metal-air Batteries*, Mills and Boon, London, 1972.

G.W. Heise and N.C. Cahoon, *The Primary Battery, Volume 1*, Wiley, New York, 1971.

G.W. Heise and N.C. Cahoon, *The Primary Battery, Volume 2*, Wiley, New York, 1976.

R. Huber, *Trockenbatterien*, Varta Aktiengesellshaft, Hanover, 1972.

R. Jasinski, *High-energy Batteries*, Plenum Press, New York, 1969.

J. Jensen, P. McGeehin and R. Dell, *Electric Batteries for Energy Storage and Conservation*, Odense University Press, Odense, 1979.

C. Julien and G.-A. Nazri, *Solid State Batteries: Materials Design and Optimization*, Kluwer Academic, 1994.

S. Kawauchi, T. Iijima and T. Kawase, *New Battery Technologies*, Matsushita Battery Industrial Co. Ltd, Osaka, 1994.

K.V. Kordesch, *Batteries: Volume 1, Manganese Dioxide*, Dekker, New York, 1974.

K.V. Kordesch, *Batteries: Volume 2, Lead Acid Batteries and Electric Vehicles*, Dekker, New York, 1977.

D. Linden, *Handbook of Batteries and Fuel Cells*, McGraw-Hill, New York, 1984.

D. Linden, *Handbook of Batteries*, 2nd edition, McGraw Hill, New York, 1995.

L.F. Martin, *Dry Cell Batteries: Chemistry and Design*, Noyes Data Corp., Park Ridge, N.J., 1973.

M.Z.A Munshi, *Handbook of Solid State Batteries & Capacitors*, World Scientific, Singapore, 1995.

D.W. Murphy, J. Broadhead and B.C.H. Steele, *Materials for Advanced Batteries*, Plenum Press, New York, 1980.

B.B. Owens, *Batteries for Implantable Biomedical Devices*, Plenum Press, New York, 1986.

G. Pistoia, *Lithium Batteries*, Elsevier, Amsterdam, 1994.

R.P. Tischer, *The Sulfur Electrode*, Academic Press, London, 1983.

C.D.S. Tuck, *Modern Battery Technology*, Ellis Horwood, New York, 1991.

G.W. Vinal, *Primary Batteries*, Wiley, New York, 1950.

G.W. Vinal, *Storage Batteries*, McGraw-Hill, New York, 1955.

E. Witte, *Blei-und Stahlakkumulatoren*, Krausskopf-Verlag, Mainz, 1967.

International power sources symposia (Brighton)

D.H. Collins, *Batteries*, 1962, Pergamon Press, Oxford, 1963.

D.H. Collins, *Batteries 2*, 1964, Pergamon Press, Oxford, 1965.

D.H. Collins, *Power Sources*, 1966, Pergamon Press, Oxford, 1967.

D.H. Collins, *Power Sources 2*, 1968, Pergamon Press, Oxford, 1969.

D.H. Collins, *Power Sources 3*, 1970, Oriel Press, Newcastle-upon-Tyne, 1971.

D.H. Collins, *Power Sources 4*, 1972, Oriel Press, Newcastle-upon-Tyne, 1973.

D.H. Collins, *Power Sources 5*, 1974, Academic Press, London, 1975.

D.H. Collins, *Power Sources 6*, 1976, Academic Press, London, 1977.

J. Thompson, *Power Sources 7*, 1978, Academic Press, London, 1979.

J. Thompson, *Power Sources 8*, 1980, Academic Press, London, 1981.

J. Thompson, *Power Sources 9*, 1982, Academic Press, London, 1983.

L.J. Pearce, *Power Sources 10*, 1984, International Power Sources Symposium Committee, 1985.

L.J. Pearce, *Power Sources 11*, 1986, International Power Sources Symposium Committee, 1986.

T. Keily and B.W. Baxter, *Power Sources 12*, 1988, International Power Sources Symposium Committee, 1989.

T. Keily and B.W. Baxter, *Power Sources 13*, 1991, International Power Sources Symposium Committee, 1991.

A. Attewell and T. Keily, *Power Sources 14*, 1993, International Power Sources Symposium Committee, 1993.

A. Attewell and T. Keily, *Power Sources 15*, 1995, International Power Sources Symposium Committee, 1995.

Series

J. O'M. Bockris and B.E. Conway, *Modern Aspects of Electrochemistry*, Volumes 1–8, Butterworth Scientific, London, 1954–74. Volumes 9–, Plenum Press, New York, 1975–.

P. Delahay and C.W. Tobias, *Advances in Electrochemistry and Electrochemical Engineering*, Volumes 1–9, Wiley, New York, 1961–1973.

H. Gerischer and C.W. Tobias, *Advances in Electrochemistry and Electrochemical Engineering*, Volumes 10–, Wiley, New York, 1977–.

A. Kozawa and K.V. Kordesch, *Progress in Batteries and Solar Cells*, JEC Press, Cleveland, 1978–.

E. Yeager and A.J. Salkind, *Techniques of Electrochemistry*, Wiley, New York, 1973–.

Reports, proceedings, etc.

Argonne National Laboratory Reports:
High Performance Batteries for EV Propulsion and Stationary Energy Storage.
Symposium and Workshop on Advanced Research and Design
Lithium/Metal Sulphide Battery Development Programme

Batteries and Fuel Cells for Stationary and Electric Vehicle Applications, 1993. (Ed. A.R. Langrebe and Z. Takehara.)

Cherry Hill Symposia:
ARL, US Army Research Laboratory, Pt Monmouth, New Jersey, USA.

DOE Report:
Status of the DOE Battery and Electrochemical Technology Programme II, R. Roberts, The Mitre Corporation, McLean, Va.

Electrochemical Society Proceedings:
Energy Storage, 1976. (Ed. J.B. Berkowitz and H.P. Silverman.)
Load Levelling, 1977. (Ed. N.P. Yao and J.R. Selman.)
Electrode Materials and Processes for Energy Conversion and Storage, 1977. (Ed. J.D.E. McIntyre, S. Srinivason and F.G. Will.)
Batteries Design and Optimisation, 1979. (Ed. S. Gross.)
Power Sources for Biomedical Implantable Applications and Ambient Temperature Lithium Batteries, 1980. (Ed. B.B. Owens and N. Margalit.)
Electrode Processes, 1979. (Ed. S. Bruckenstein, J. McIntyre, B. Miller and E. Yeager.)
Lithium Non-aqueous Battery Electrochemistry, 1980. (Ed. E. Yeager, B. Schumm, G. Blomgren, D. Blankenship, V. Leger and J. Akridge.)
Lithium Batteries, 1981. (Ed. B. Venkatasetty.)
The Nickel Electrode, 1982. (Ed. R. Gunther and S. Gross.)
Corrosion in Batteries and Fuel Cells and Corrosion in Solar Energy Systems, 1983. (Ed. C. Johnson and S. Pottmen.)
Membranes and Ionic and Electronic Conducting Polymers, 1983. (Ed. E. Yeager, K. Mauritz, D. Blankenship, B. Schumm, K. Abbey and J. Ackridge.)
Lithium Batteries, 1984. (Ed. A. Dey.)

Advances in Battery Materials and Process, 1984. (Ed. J. McBreen, R. Yeo, D. Chin and A. Tseng.)

Porous Electrodes: Theory and Practice, 1984. (Ed. H. Maru, T. Katan and M. Klein.)

Advances in Lead–Acid Batteries, 1984. (Ed. K. Bullock and D. Pavlov.)

Manganese Dioxide Electrode Theory and Practice for Electrochemical Applications, 1985. (Ed. B. Schumm, R. Middaugh, M. Grotheer and J. Hunter.)

Electrochemical and Thermal Modeling of Battery, Fuel Cell and Photoenergy Conversion Systems, 1986. (Ed. J. Selman and H. Maru.)

Lithium Batteries, 1987. (Ed. A. Dey.)

Sodium Sulfur Battery, 1987. (Ed. R. Weaver and R. Sen.)

Second Symposium on Electrode Materials and Processes for Energy Conversion and Storage, 1987. (Ed. S. Srinavasan, S. Wagner and H. Wroblowa.)

High Voltage and Smart Power Devices, 1987. (Ed. P. Schalke.)

History of Battery Technology, 1987. (Ed. A. Salkind.)

Second Annual Battery Conference on Applications and advances. (Ed. R. Das, B. Eliash, H. Fran and D. Picket.)

Primary and Secondary Ambient Temperature Lithium Batteries, 1988. (Ed. J. Gabano, Z. Takebara and P. Bro.)

Stationary Energy Storage: Load Levelling and Remote Applications, 1988. (Ed. A. Landrebe, D. Rand, S. Voorhees and R. Sen.)

Materials and Processes for Lithium Batteries, 1989. (Ed. K. Abraham and B. Owens.)

High Voltage and Smart Power ICs, 1989. (Ed. M. Shibib.)

Nickel Hydroxide Electrodes, 1990. (Ed. D. Corrigan and A. Zimmerman.)

Rechargeable Lithium Batteries, 1990. (Ed. S. Subbarao, V. Koch, B. Owens and W. Smyrle.)

Primary and Secondary Lithium Batteries, 1991. (Ed. M. Soloman.)

Modelling of Batteries and Fuel Cells, 1991. (Ed. M. Verbrugge and J. Stockel.)

Hydrogen Storage Materials, Batteries, and Electrochemistry, 1992. (Ed. D. Corrigan and S. Srinavasan.)

High Power, Ambient Temperature and Lithium Batteries, 1992. (Ed. W. Clark and G. Halpert.)

Microscopic Models of Electrode/Electrolyte Interfaces, 1993. (Ed. J. Halley and L. Blum.)

Batteries and Fuel Cells for Stationary and Electric Vehicles, 1993. (Ed. A. Landgrebe and Z. Takehara.)

New Sealed Rechargeable Batteries and Supercapacitors, 1993. (Ed. B. Barnett, E. Dowgiallo, G. Halperi, Y. Matsude and Z. Takehara.)

Lithium Batteries IV, 1993. (Ed. B. Surumpudi and V. Kock.)

Lithium Batteries, V, 1994. (Ed. N. Doddapaneni and A. Langrebe.)

Electrode Materials and Processes for Energy Conversion and Storage, 1994. (Ed. S. Srinavasan, D.D. Macdonald and A.C. Khandkar.)

Hydrogen and Metal Hydride Batteries, 1994. (Ed. P.D. Bennet and T. Sakai.)

Rechargeable Zinc Batteries (Frumkin Symposium), 1995. (Ed. A.J. Salkind, F. MacLarnon and V.S. Bagotzky.)

Supercapacitors, 1995. (Ed. F.M. Delnick and Tomkiewicz.)

Electrode Process, 1996. (Ed. A. Wiekowski and K. Itaya.)

Exploratory Research and Development of Batteries for Electric and Hybrid Vehicles, 1996. (Ed. W. Adams, A., Landgrebe and B. Scrosati.)

Aqueous Batteries, 1996. (Ed. P. Bennet and J. Gross.)
Lithium Polymer Batteries, 1996. (Ed. J. Broadhead and B. Scrosati.)
Electrochemical Capacitors, 1996. (Ed. F.M. Delnick.)

Exploratory Research and Development of Batteries for Electric and Hybrid Vehicles, 1996. (Ed. W.A. Adams, A.R. Landgrebe and B. Scrosati).

House of Lords Select Committee on Science and Technology: Report:
Electric Vehicles, H.M. Stationary Office, London, 1980.

Lawrence Livermore Laboratory (UCRL - 52553):
Energy Storage Systems for Automobile Propulsion, 1978. (E. Behrin).

New Technology for the Treatment of Used Dry Batteries, Sumitomo Heavy Industries Ltd, Bulletin, Revision 6, June 1991.

Proceedings of International Lithium Battery Meetings (IMLB):
IMLB-2: *J. Power Sources*, 14, 1985.
IMLB-3: *J. Power Sources*, 20, 1985.
IMLB-4: *J. Power Sources*, 26, 1985.
IMLB-6: *J. Power Sources*, 43/44, 1985.
IMLB-7: *J. Power Sources*, 54, 1985.
IMLB-8: *J. Power Sources*, 1997, in press.

Proceedings of Power Source Conferences, New Jersey, USA, 1970–96.

Journals

Electrochimica Acta, International Society for Electrochemistry, Pergamon Press, Oxford.
EPRI Journal.
Journal of Applied Electrochemistry, Chapman and Hall, London.
Journal of Electroanalytical Chemistry, Elsevier Sequoia S.A., Lausanne.
Journal of the Electrochemical Society, The Electrochemical Society, Princeton, N.J.
Journal of Power Sources, Elsevier Sequoia, S.A., Lausanne.
Solid State Ionics, North-Holland, Amsterdam.

Glossary

activation overpotential (overvoltage) Contribution to the total overpotential due to the charge transfer step at the electrode interface.

active mass The material in an electrochemical cell which takes part in the cell reaction. For example, the lead oxide contained in the positive plate of a lead–acid battery.

anode The electrode at which oxidation takes place and which gives up electrons to the external circuit.

anolyte The electrolytic phase in contact with the anode.

battery An assembly of two or more cells electrically connected to form a unit. For example, a 12 V SLI battery is made up of six 2 V cells in series. However, the term is also often used to indicate a single cell.

binder A material, usually polymeric, added to the active mass to increase its mechanical strength.

bipolar electrode An electrode assembly which functions as the anode of one cell on one side, and as the cathode of the next cell on the other side. Also known as a 'duplex' electrode, especially in Leclanché batteries.

bus; bus bar A rigid metallic conductor which connects different elements of a battery; also, the conductor for an electrical system to which a battery terminal is attached.

button cell Miniature cylindrical cell having a characteristic disc shape.

can; case The external envelope of a cell or battery, or the box containing the cells and connectors.

capacity, rated The value of the output capability of a battery, expressed in Ah, at a given discharge rate before the voltage falls below a given cut-off value, as indicated by the manufacturer.

cathode The electrode at which reduction takes place and which withdraws electrons from the external circuit.

catholyte The electrolytic phase in contact with the cathode.

cell Electrochemical device which directly interconverts chemical and electrical energy.

cell reversal Inversion of the polarity of the terminals of a cell in a multicell battery. Cell reversal is usually due to overdischarge, when differences in the capacity of individual cells result in one or more cells reaching complete discharge before the others.

charge acceptance The ability of a secondary cell or battery to convert the active material to a dischargeable form. It is measured by the capacity which can be subsequently delivered to a load as a result of the charging process. If the charge acceptance is 100% then all of the electrical energy input would become available for useful output.

charge rate See: C-rate.

charge retention The ability of a charged cell to resist self-discharge.

charge, state of The condition of a cell or a battery in terms of the remaining available capacity.

collector; current collector Electronic conductor embedded in the active mass and connected to the bus bar or terminal.

concentration overpotential (overvoltage) Contribution to the total overpotential due to non-uniform concentrations in the electrolyte phase near the electrode surface caused by the passage of current.

corrosion Oxidation of a metallic phase starting at the surface and caused by the reaction of the metal with components of the environment. In batteries, corrosion phenomena play an important role, especially in the case of primary aqueous cells and in high temperature systems.

C-rate A method for expressing the rate of charge or discharge of a cell or battery. A cell discharging at a C-rate of τ will deliver its nominal rated capacity in $1/\tau$ h; e.g. if the rated capacity is 2 Ah, a discharge rate of C/1 corresponds to a discharge current of 2 A, a rate of C/10 to 0.2 A, etc.

creep The process by which liquid electrolytes, and in particular alkalis, can escape past rubber–metal or polymer–metal seals, or through minute cracks in a cell case or lid.

current density Electric flux per unit area. It is generally defined in terms of the geometric or projected electrode area and is measured in A/m^2 or mA/cm^2.

cut-off voltage Final voltage of a discharge or charge operation. In the case of discharge, it is chosen as the voltage value below which the connected equipment will not operate, or below which operation is not recommended because of the onset of irreversible processes in the cell. In the case of charge, it is selected to allow complete conversion of active material with a minimum of gassing or other overcharge reactions.

cycle Sequence of charge and discharge of a secondary battery.

cycle life The total number of charge/discharge cycles that can be delivered by a secondary cell or battery, while maintaining a predetermined output capacity and cycle energy efficiency.

depolarizer A substance which is supposed to reduce electrode polarization. The term was introduced when it was believed that the electrode polarization was due solely to gas evolution at the electrode and the action of the depolarizer was to eliminate or prevent this process. Today views are different, but the term is still sometimes used, usually to denote the positive active material.

discharge curve A plot of cell or battery voltage as a function of time or of discharge capacity, under a defined discharge current or load.

discharge depth The percentage of the capacity to which a cell or battery has been discharged. Shallow/deep discharge: small/large fraction of the usable capacity consumed.

discharge rate See 'C-rate'.

drain Withdrawal of current from a cell or battery.

dry cell A cell in which the electrolyte is immobilized, being either in the form of a paste or gel or absorbed in a microporous separator material.

dry charged cell A cell which is in its fully charged state but without electrolyte.

duplex electrode Type of electrode system used in flat Leclanché multi-cell batteries, formed by zinc coated on one side with carbon. It acts as the cathode current collector for one cell and as the anode for the adjacent cell. (see 'bipolar electrode').

electrode The electronic conductor and associated active materials at which an electrochemical reaction occurs.

electrodeposition Deposition of a chemical species at the electrode of an electrolytic cell caused by the passage of electric current.

electrolysis Chemical modifications (i.e. oxidation and reduction) produced by passing an electric current through an electrolyte.

electrolyte The medium which permits ionic conduction between positive and negative electrodes of a cell. It may be solid or liquid. In some cases the electrolyte may take part in the cell reaction.

equalizing charge Passage of an amount of charge by which the under-charged cells of a battery are brought up to a fully charged condition without damaging those already fully charged.

expander A substance added in small amount to the active materials of a lead–acid battery to improve the service life and capacity of the electrodes. In particular, an expander prevents the increase in crystal grain size of lead in the negative electrode.

failure The state in which the performance of a cell or battery does not meet the normal specifications.

float charging Method of recharging in which a secondary battery is continuously connected to a constant voltage supply that maintains the cell in fully charged condition.

forming; formation A series of charge/discharge cycles carried out under carefully controlled conditions after the manufacture of a secondary cell in order to optimize the morphology of the active mass.

fuel cell An electrochemical generator in which the reactants are stored externally and may be supplied continuously to the cell.

gassing Gas evolution which takes place towards the end of the charging of a battery.

grain boundary The surface separating two regions of a solid having different crystal orientations.

grid The lead framework of a lead–acid battery plate which holds the active material in place.

group A set of electrodes within a cell which are connected in parallel.

hybrid cell Electrochemical cell in which *one* of the two active reagents is in the gas phase and may be supplied from an external source. A hybrid cell occupies an intermediate position between closed cells and fuel cells.

hybrid electric vehicle A vehicle that has more than one type of power supply to support the drive, e.g. battery/motor plus fossil fuel/internal combustion engine.

immobilized electrolyte See 'dry cell'.

inhibitor A substance added to the electrolyte which prevents an electrochemical process, generally by modifying the surface state of an electrode. A well known example is that of corrosion inhibitors which prevent metal corrosion.

initial drain Current that a cell or battery supplies when first placed on a fixed load.

internal resistance Resistance to the flow of direct current within a cell, causing a drop in closed circuit voltage proportional to the current drain from the cell.

iR loss; iR drop Decrease in the voltage of a cell during the passage of current due to the internal resistance of the bulk phases within the cell – mainly that of the electrolyte and the separators. Also known as 'ohmic loss'.

lithium ion cell A secondary lithium cell in which both the negative and positive electrodes are lithium insertion compounds; also known as rocking chair, shuttlecock or swing cell.

load The external devices or circuit elements to which electric power is delivered by a cell or battery.

load levelling The intervention aimed at reducing non-uniform conditions in electricity demand. The principle of load levelling is to store energy when demand is low and to use it to meet peak demand.

loss See '*iR* loss'; 'polarization loss'.

LPB (lithium polymer battery) A cell (generally rechargeable) having a lithium foil negative, a metal oxide positive and a polymer electrolyte.

maintenance The procedures which are required in order to keep a battery in proper operating conditions. They may include trickle-charging to compensate for self-discharge, addition of water to the electrolyte, etc.

mass transport Transfer of materials consumed or formed in an electrode process to or from the electrode surface. Mechanisms of mass transport may include diffusion, convection and electromigration.

memory effect Process occurring in nickel–cadmium and some nickel–metal hydride cells which causes a temporary reduction in capacity following repetitive shallow charge–discharge cycles.

negative Negatively charged electrode, usually of a secondary cell; acts as anode during discharge and cathode during charge.

ohmic loss See '*iR* loss'.

open circuit voltage The voltage of a cell or battery under no-load condition, measured with a high impedance voltmeter or potentiometer.

overcharge The continued application of charging current to a cell or battery after it has reached its maximum state of charge.

overdischarge Forced discharge of a cell or a battery past 100% of the available capacity. In the case of a multicell battery, the overdischarge may cause cell reversal.

overpotential; overvoltage Difference between the actual electrode voltage when a current is passing and the equilibrium (zero current) potential. A number of different effects may contribute to the total overvoltage.

oxidation The loss of electrons by a chemical species.

passivation Surface modifications of metallic materials which cause an increase in their resistance to corrosion process.

passivity The condition of a metallic material corresponding to an immeasurably small rate of corrosion.

plate In the terminology of secondary batteries, this has the same meaning as 'electrode'.

PLI battery (plastic lithium ion battery) A lithium ion cell which uses a polymer electrolyte.

polarization Deviation from equilibrium conditions in an electrode or galvanic cell caused by the passage of current. It is related to the irreversible phenomena at the electrodes (electrode polarization) or in the electrolytic phase (concentration polarization).

polarization loss Reduction in the voltage of a cell delivering current from its equilibrium value.

polymer electrolyte A macroscopically solid electrolyte based either on salts dissolved in a cation-coordinating polymer such as poly(ethylene oxide), or on a gel formed by a salt, organic solvent(s) and a structural polymer such as poly(acrylonitrile).

positive Positively charged electrode, usually of a secondary cell; acts as cathode during discharge and anode during charge.

post See 'terminal'.

primary battery A cell or battery whose useful life is over once its reactants have been consumed, i.e. one not designed to be recharged.

rate See 'C-rate'.

recombining cell A secondary cell in which provision has been made for the products of overcharge reactions to recombine so that no net change occurs to the composition of the cell system as a result of overcharging.

reduction The gain of electrons by a chemical species.

reserve battery In principle, any battery which will not deliver current in its manufactured form until activated by a suitable procedure, e.g. by adding the electrolyte to the dry components (water activated cells or cells activated by addition of special electrolytes), or by raising the temperature of the cell (thermal batteries, where the electrolyte is generally a mixture of salts in the solid state at ambient temperature).

reversal See 'cell reversal'.

rocking chair cell A lithium ion cell.

secondary battery; storage battery Cell or battery which can be recharged after discharge, under specific conditions.

self-discharge Capacity loss of a cell or battery under open circuit conditions due to chemical reactions within the cell.

self-discharge rate The rate at which a cell or battery loses service capacity when standing idle.

separator Electrically insulating layer of material which physically separates electrodes of opposite polarity. Separators must be permeable to the ions of the electrolyte and may also have the function of storing or immobilizing the electrolyte.

service life Timescale of satisfactory performance of a battery under a specified operating schedule, expressed in units of time or number of charge/discharge cycles.

shedding The process whereby poorly adhering active mass (generally in the positive plate of a lead–acid cell) falls from the grid to form a sludge (mud) on the floor of the cell.

shelf life Period of time a cell can be kept idle after manufacture without significant deterioration.

short circuit The condition when the terminals of a cell or battery are connected directly.

shuttlecock cell A lithium ion cell.

SLI battery A battery of usually 12 or 24 V used for starting, lighting and ignition in vehicles with internal combustion engines.

stack An assembly of parallel plates.

storage battery See 'secondary battery'.

surface-active agent A substance which modifies the behaviour of a phase by interacting with its surface. For example, in the case of lead–acid batteries, the morphology of the active materials deposited at the electrodes may be strongly affected by the addition of surface-active agents.

swing cell A lithium ion cell.

terminal The external electric connections of a cell or battery; also known as 'terminal post' or 'post'.

thermal battery A type of reserve cell which is activated by raising the temperature.

thermal management The means whereby a battery system is maintained within a specified temperature range while undergoing charge or discharge.

thermal runaway A process in which a cell undergoes an uncontrolled rise in temperature due to the passage of increasing current (on, say, short-circuit discharge or constant voltage charging) as the temperature rises.

trickle charging Method of recharging in which a secondary cell is either continuously or intermittently connected to a constant current supply in order to maintain the cell in fully or nearly fully charged condition.

uninterruptable power supply (UPS) A power system which maintains current flow without even a momentary break, in the event of mains or generator failure.

vent Valve mechanism which allows controlled escape of gases generated during charging, but prevents spillage of electrolyte.

voltage delay Time interval at the start of a discharge during which the working voltage of a cell is below its steady value. The phenomenon is generally due to the presence of passivating films on the negative electrode.

wet cell A cell in which the liquid electrolyte is free-flowing.

Index

Note: 'n.' after a page number indicates a footnote on that page.

Printed and bound by CPI Group (UK) Ltd, Croydon, CR0 4YY

03/10/2024

01040432-0016